Springer Theses

Recognizing Outstanding Ph.D. Research

For further volumes:
http://www.springer.com/series/8790

Aims and Scope

The series "Springer Theses" brings together a selection of the very best Ph.D. theses from around the world and across the physical sciences. Nominated and endorsed by two recognized specialists, each published volume has been selected for its scientific excellence and the high impact of its contents for the pertinent field of research. For greater accessibility to non-specialists, the published versions include an extended introduction, as well as a foreword by the student's supervisor explaining the special relevance of the work for the field. As a whole, the series will provide a valuable resource both for newcomers to the research fields described, and for other scientists seeking detailed background information on special questions. Finally, it provides an accredited documentation of the valuable contributions made by today's younger generation of scientists.

Theses are accepted into the series by invited nomination only and must fulfill all of the following criteria

- They must be written in good English.
- The topic of should fall within the confines of Chemistry, Physics and related interdisciplinary fields such as Materials, Nanoscience, Chemical Engineering, Complex Systems and Biophysics.
- The work reported in the thesis must represent a significant scientific advance.
- If the thesis includes previously published material, permission to reproduce this must be gained from the respective copyright holder.
- They must have been examined and passed during the 12 months prior to nomination.
- Each thesis should include a foreword by the supervisor outlining the significance of its content.
- The theses should have a clearly defined structure including an introduction accessible to scientists not expert in that particular field.

Bernd Terhalle

Controlling Light in Optically Induced Photonic Lattices

Doctoral Thesis accepted by
Westfälische Wilhelms-Universität Münster, Germany

 Springer

Author
Dr. Bernd Terhalle
Laboratory for Micro-
    and Nanotechnology
Paul Scherrer Institut
Villigen PSI
ODRA 113
Switzerland
e-mail: bernd.terhalle@psi.ch

Supervisor
Prof. Dr. Cornelia Denz
Institut für Angewandte Physik
Westfälische Wilhelms-Universität Münster
Corrensstr. 2/4
48149 Münster
Germany
e-mail: denz@uni-muenster.de

ISSN 2190-5053   e-ISSN 2190-5061

ISBN 978-3-642-26702-4   ISBN 978-3-642-16647-1 (eBook)

DOI 10.1007/978-3-642-16647-1

Springer Heidelberg Dordrecht London New York

© Springer-Verlag Berlin Heidelberg 2011
Softcover reprint of the hardcover 1st edition 2011

This work is subject to copyright. All rights are reserved, whether the whole or part of the material is concerned, specifically the rights of translation, reprinting, reuse of illustrations, recitation, broadcasting, reproduction on microfilm or in any other way, and storage in data banks. Duplication of this publication or parts thereof is permitted only under the provisions of the German Copyright Law of September 9, 1965, in its current version, and permission for use must always be obtained from Springer. Violations are liable to prosecution under the German Copyright Law.

The use of general descriptive names, registered names, trademarks, etc. in this publication does not imply, even in the absence of a specific statement, that such names are exempt from the relevant protective laws and regulations and therefore free for general use.

Cover design: eStudio Calamar, Berlin/Figueres

Printed on acid-free paper

Springer is part of Springer Science+Business Media (www.springer.com)

Supervisor's Foreword

Discrete periodic structures play an important role in physics, and have revealed an exciting new area in photonics in the last years. It is not only the change in the laws of electromagnetic wave propagation that gives rise to a completely new behavior, but also the generality of this behavior appearing in solid state physics, optics or in cold atoms that makes this field so highly attractive for investigations of basic quantum and nonlinear physics phenomena. Also, periodic structures play a more and more important role in applications as there are photonic crystals, metamaterials or cold atoms in lattices. Nowadays, questions how to control light in these periodic structures in two or three dimensions represent still a challenge, especially, because strong nonlinearities operate in the regime of discrete periodicity. It is this highly actual field that is addressed in the present thesis.

Using the model system of a photorefractive nonlinearity that allows to create photonic lattices by light and control them at the same time with light, the thesis presents a comprehensive picture of nonlinear and quantum optics phenomena in photonic lattices that have immense consequences for applications in optical information processing.

In a first part, the similarities between light propagation in periodic refractive index structures and electrons in crystalline solids are exploited to understand and experimentally realize fundamental phenomena of solid state quantum physics in optics: Bloch oscillations and Zener tunneling. Both are associated with the propagation of waves or quantum particles in periodic potentials under the action of an external driving force and were originally predicted for electrons moving in a periodic potential with a superimposed constant electric field. In this situation, the particles do not just follow the driving force but perform an oscillatory motion known as Bloch oscillation. However, these oscillations do not persist forever, but are damped by interband transitions known as Zener tunneling. Optical analogs of Bloch oscillations and Zener tunneling are demonstrated in this thesis for the first time in a two-dimensional hexagonal lattice. In such an optical setting, the periodic modulation of the refractive index plays the role of the crystalline potential, while an additional linear refractive index ramp acts as external force. Studying the threefold resonance of the hexagonal structure, the thesis shows symmetric tunneling as well as asymmetric tunneling or tunneling of Rabi oscillations. Furthermore, it is shown that the observed effects can be employed for the controlled generation of Bloch waves and thus the characterization of photonic structures. These results are corroborated by numerical simulations using an anisotropic photorefractive model description.

The second part deals with one of the most spectacular fields of nonlinear dynamics of coherent light and matter waves in periodic potentials—the properties of vortices and vortex flows in optical lattices. Dramatic changes of light diffraction or tunneling of matter waves in media with periodically modulated parameters offer novel directions for manipulating waves with a complex phase structure. In optics, self-trapped phase singularities in the form of isolated discrete vortices have been realized in simple one- and two-dimensional photonic lattices already a couple of years ago. However, spatially multi vortex coherent states remained unobserved and largely unexplored since a couple of years due to their complex stability conditions depending on the nature of the nonlinearity. The present thesis has resolved this challenge and includes demonstration of the existence of stable multi vortex solitons as well as soliton clusters for the first time in an optical system.

It is the overall understanding of light propagation in complex photonic lattices—combining theoretical description, numerical simulations, and impressive experimental results—that makes this thesis a milestone in complex beam propagation in nonlinear photonics lattices. With 9 publications in highly ranked international journals and two book chapters, the work of Bernd Terhalle that has been performed between 2007 and 2010 at the Nonlinear Photonics Group and the Center for Nonlinear Science at the University of Münster, Germany, as well as at the Nonlinear Physics Centre at the Australian National University in Canberra, Australia, is internationally highly recognized, and underlines the comprehensive view of linear and nonlinear propagation of complex light fields in discrete periodic structures that is given in the thesis.Therefore, the thesis was honored summa cum laude by the University of Münster in 2010.

Münster, September 2010   Prof. Dr. Cornelia Denz

Acknowledgments

The present thesis is the result of different research projects undertaken at the Westfälische Wilhelms-Universität in Münster and the Australian National University in Canberra. As research is teamwork, various people contributed in different ways to the completion of this work and it is my pleasure to express my sincerest gratitude at this point.

First of all, I thank Prof. Dr. Cornelia Denz for continuous guidance of my scientific work and for providing help, inspiration and encouragement whenever needed. Her support of my scholarship application and the opportunity to present our results on different national as well as international conferences have made the last three years a great experience to me. I also thank Prof. Dr. Yuri Kivshar who kindly agreed to serve as second supervisor in the binational PhD project and gave me the opportunity to work in his group at the Australian National University in Canberra for several months each year.

I am deeply indepted to Dr. Anton Desyatnikov for excellent supervison throughout the whole PhD project and especially during my stays in Canberra. His support in all scientific as well as non-scientific issues, his enthusiasm, and most importantly his endless patience in explaining theory to an experimentalist have been invalueable for the completion of this thesis.

Special thanks go to Dr. Tobias Richter for great collaboration with inspiring ideas and strong numerical support, even after finishing his own PhD thesis.

I am grateful to Patrick Rose for working together with me in the field of optically induced photonic lattices in a very cooperative and productive working atmosphere with many enlightening discussions and exciting results. Furthermore, I thank him for proofreading the manuscript.

Additional thanks go to my former diploma student Dennis Göries who significantly contributed to this work with his experiments on vortex solitons in optically induced lattices. For supporting my experiments in Canberra, I would especially like to thank Prof. Dr. Wieslaw Krolikowski and Dr. Dragomir Neshev.

Also, I thank all present and former members of both working groups who have accompanied my work in Münster and Canberra. In particular, I would like to mention Dr. Jörg Imbrock, Wolfgang Horn, Artur Davoyan, Aliaksandr Minovich, and Ksawery Kalinowski. My deepest thanks go to my parents and my family for unconditional support and encouragement in any possible situation.

Above all, however, I wish to thank you, Anne, for all your love, support and understanding and especially for standing by me while being separated thousands of miles. Thank you so much!

Bernd Terhalle

Contents

Chapter 1
Motivation and Outline

In 1961, shortly after the presentation of the first working laser by Maiman [1], Franken and coworkers performed an experiment that considerably changed the understanding of optics at that time [2]. They focused a ruby laser at a wavelength of 694.3 nm into a crystalline quartz plate and demonstrated that the emergent spectrum contained radiation at twice the input frequency, i.e. at a wavelength of 347.15 nm. This first observation of so-called second harmonic generation is nowadays widely accepted as the starting point of the field of *nonlinear optics* which up to now has developed into a very active research area ranging from fundamental studies of light-matter interactions to a variety of optical applications [3].

Prior to 1961, the propagation of light was mostly regarded as a linear phenomenon and all experiments indicated that transmission, reflection and refraction in transparent materials were affected neither by the intensity of the light nor by the presence of other beams [4]. In this regime of *linear optics*, the principle of superposition applies and the light intensity at the output is always directly proportional to the input intensity. Furthermore, the frequencies at the input and the output are always the same.

However, if the electric field of an intense light wave, such as a laser beam, becomes comparable to the intra-atomic field, the interaction between the light and the optical material through which it propagates results in a local modification of the polarization density and the refractive index. This is the regime of nonlinear optics in which an optical beam can dynamically affect its own propagation as well as the propagation of other beams. Therefore, it offers great potential for the realization of efficient and dynamic tools to control light by light itself for future all-optical technologies.

Another major breakthrough in the context of controlling light propagation was made in 1987 when Yablonovitch [5] and John [6] independently suggested the use of dielectric structures with spatially modulated refractive indices. It has been shown that wave propagation in these structures is governed by the concept of photonic band gaps which can be seen as an optical analog of the electronic band gaps in crystalline solids where the periodic potential of the atoms leads to forbidden regions in the transmission spectrum.

B. Terhalle, *Controlling Light in Optically Induced Photonic Lattices*, Springer Theses, DOI: 10.1007/978-3-642-16647-1_1, © Springer-Verlag Berlin Heidelberg 2011

Due to this analogy, periodically modulated refractive index structures are nowadays commonly denoted as *photonic crystals* with examples ranging from simple one-dimensional Bragg gratings and waveguide arrays [7] over two-dimensional photonic crystal fibers [8] to more complex three-dimensional photonic crystals which are known to exist in different geometries [9]. The important common feature of all these structures is the ability to manipulate the flow of light in the direction of periodicity. Up to now, several approaches for controlling light propagation in photonic crystals have been suggested theoretically and demonstrated experimentally. Examples include the transmission of light around sharp corners [10] as well as more sophisticated effects like negative refraction [11] or self-collimation [12].

To increase the control possibilities even further, the effects of nonlinearity and periodicity can be combined in *nonlinear photonic crystals* [13]. In this case, the nonlinear response of the material allows for dynamic tunability of the structures by varying the light intensity and the interplay between nonlinearity and periodicity is exploited to efficiently manipulate light by light for optical switching and signal processing applications. An important example is given by the *discrete soliton* [14] as a localized wavepacket arising from the balance between diffraction (or dispersion) and nonlinearity which has been observed in both one- and two-dimensional geometries [14, 15].

The main objective of this thesis is the experimental investigation of novel schemes for light control based on fundamental nonlinear wave physics in periodic media. For this reason, an experimental platform is required that allows for studying periodic structures of different geometries which are easy to fabricate, enable a complete control over structural parameters and possess strong nonlinearities, preferably at low laser powers.

A genuine way to fulfill all these requirements is to create a refractive index grating by interfering two or more coherent laser beams inside a photorefractive material, typically Strontium Barium Niobate (SBN). Such materials enable wavelength-sensitive and reconfigurable nonlinear refractive index patterns which can be induced at very low power levels. Thus, they offer an ideal test bench for the investigation of fundamental phenomena of wave propagation in periodic media and have consequently been employed for the experimental demonstration of various effects [16–18]. For the same reason, optically induced photonic structures in photorefractive SBN crystals form the basis of all experiments presented in this thesis.

The structure of the thesis is as follows. Starting from Maxwell's equations, Chap. 2 provides a general mathematical introduction to wave propagation in nonlinear photonic media resulting in a scalar propagation equation for the amplitude of the electric field. The photorefractive nonlinearity as a key mechanism of the presented work is explained in detail and a realistic, anisotropic model for the induced refractive index change is derived. Finally, the propagation equation is modified to include a periodic refractive index modulation and the concept of photonic band gaps is introduced.

A detailed description of optically induced photonic lattices in photorefractive media is given in Chap. 3 [19–22]. The use of nondiffracting waves to create two-dimensional, transversally periodic photonic structures is explained and different lattice geometries are discussed. In particular, it is shown that, due to the photorefractive anisotropy, the resulting refractive index structure strongly depends on the spatial orientation of the lattice wave and may exhibit significant symmetry reductions. Different ways to minimize the effects of anisotropy are discussed as well.

Chapter 4 shows the possibility of resonant transitions between the high-symmetry points of the Brillouin zone. It is demonstrated that the one-dimensional Landau–Zener–Majorana model describing interband transitions in solid-state physics can be extended to the two-dimensional case of hexagonal photonic structures. Based on this model, Rabi oscillations and Landau–Zener tunneling are observed in experiments and the results are compared to direct numerical simulations using the anisotropic photorefractive model [23]. Finally, the influence of nonlinearity on the observed transitions is investigated.

The stabilization of complex light fields by nonlinear light localization is the main objective of Chap. 5. Starting from a general introduction to solitons in periodic photonic structures, the focus of this chapter is put on the existence and stability of soliton clusters with a superimposed phase structure containing one or more phase singularities (optical vortices). In particular, it is shown that the photorefractive anisotropy can be employed to control the vortex stability [24] and the existence of stable ring-shaped discrete vortex solitons [25] as well as multi-vortex solitons [26] is demonstrated.

The thesis then concludes with a summary of the presented results and an outlook on possible investigations that may be affiliated to this work.

References

1. Maiman, T.H.: Stimulated optical radiation in ruby. Nature **187**, 493 (1960)
2. Franken, P.A., Hill, A.E., Peters, C.W., Weinreich, G.: Generation of optical harmonics. Phys. Rev. Lett. **7**, 118 (1961)
3. Boyd, R.W.: Nonlinear Optics, 3rd edn. Academic Press, New York (2008)
4. Yeh, P.: Introduction to Photorefractive Nonlinear Optics. Wiley, New York (1993)
5. Yablonovitch, E.: Inhibited spontaneous emission in solid-state-physics and electronics. Phys. Rev. Lett. **58**, 2059 (1987)
6. John, S.: Strong localization of photons in certain disordered dielectric superlattices. Phys. Rev. Lett. **58**, 2486 (1987)
7. Yeh, P.: Optical Waves in Layered Media. Wiley, New York (1998)
8. Russell, P.S.J.: Photonic crystal fibers. Science **299**, 358 (2003)
9. Joannopoulus, J.D., Meade, R.D., Winn, J.N.: Photonic Crystals: Molding the Flow of Light. Princeton University Press, Princeton (1995)
10. Mekis, A., Chen, J.C., Kurland, I., Fan, S., Villeneuve, P.R., Joannopoulos, J.D.: High transmission through sharp bends in photonic crystal waveguides. Phys. Rev. Lett. **77**, 3787 (1996)

11. Cubukcu, E., Aydin, K., Ozbay, E., Foteinopoulou, S., Soukoulis, C.M.: Electromagnetic waves: negative refraction by photonic crystals. Nature **423**, 604 (2003)
12. Kosaka, H., Kawashima, T., Tomita, A., Notomi, M., Tamamura, T., Sato, T., Kawakami, S.: Self-collimating phenomena in photonic crystals. Appl. Phys. Lett. **74**, 1212 (1999)
13. Mingaleev, S.F., Kivshar, Y.S.: Nonlinear photonic crystals: toward all-optical technologies. Opt. Photon. News **13**, 48 (2002)
14. Eisenberg, H.S., Silberberg, Y., Morandotti, R., Boyd, A.R., Aitchison, J.S.: Discrete spatial optical solitons in waveguide arrays. Phys. Rev. Lett. **81**, 3383 (1998)
15. Fleischer, J.W., Segev, M., Efremidis, N.K., Christodoulides, D.N.: Observation of two-dimensional discrete solitons in optically induced nonlinear photonic lattices. Nature **422**, 147 (2003)
16. Fleischer, J., Bartal, G., Cohen, O., Schwartz, T., Manela, O., Freedman, B., Segev, M., Buljan, H., Efremidis, N.: Spatial photonics in nonlinear waveguide arrays. Opt. Express **13**, 1780 (2005)
17. Neshev, D.N., Sukhorukov, A.A., Krolikowski, W., Kivshar, Y.S.: Nonlinear optics and light localization in periodic photonic lattices. J. Nonlinear Opt. Phys. Mater. **16**, 1 (2007)
18. Lederer, F., Stegeman, G., Christodoulides, D., Assanto, G., Segev, M., Silberberg, Y.: Discrete solitons in optics. Phys. Rep. **463**, 1 (2008)
19. Desyatnikov, A.S., Sagemerten, N., Fischer, R., Terhalle, B., Träger, D., Neshev, D.N., Dreischuh, A., Denz, C., Krolikowski, W., Kivshar, Y.S.: Two-dimensional self-trapped nonlinear photonic lattices. Opt. Express **14**, 2851 (2006)
20. Terhalle, B., Trager, D., Tang, L., Imbrock, J., Denz, C.: Structure analysis of two-dimensional nonlinear self-trapped photonic lattices in anisotropic photorefractive media. Phys. Rev. E **74**, 057601 (2006)
21. Terhalle, B., Desyatnikov, A.S., Bersch, C., Träger, D., Tang, L., Imbrock, J., Kivshar, Y.S., Denz, C.: Anisotropic photonic lattices and discrete solitons in photorefractive media. Appl. Phys. B **86**, 399 (2007)
22. Rose, P., Richter, T., Terhalle, B., Imbrock, J., Kaiser, F., Denz, C.: Discrete and dipole-mode gap solitons in higher-order nonlinear photonic lattices. Appl. Phys. B **89**, 521 (2007)
23. Terhalle, B., Desyatnikov, A.S., Neshev, D.N., Krolikowski, W., Denz, C., Kivshar, Y.S.: Observation of Landau–Zener Tunneling in Hexagonal Photonic Lattices. Photorefractive Materials, Effects and Devices, ISBN 978-3-00-027892-1 (2009)
24. Terhalle, B., Göries, D., Richter, T., Rose, P., Desyatnikov, A.S., Kaiser, F., Denz, C.: Anisotropy-controlled topological stability of discrete vortex solitons in optically induced photonic lattices. Opt. Lett. **35**, 604 (2010)
25. Terhalle, B., Richter, T., Law, K.J.H., Göries, D., Rose, P., Alexander, T.J., Kevrekidis, P.G., Desyatnikov, A.S., Krolikowski, W., Kaiser, F., Denz, C., Kivshar, Y.S.: Observation of double-charge discrete vortex solitons in hexagonal photonic lattices. Phys. Rev. A **79**, 043821 (2009)
26. Terhalle, B., Richter, T., Desyatnikov, A.S., Neshev, D.N., Krolikowski, W., Kaiser, F., Denz, C., Kivshar, Y.S.: Observation of multivortex solitons in photonic lattices. Phys. Rev. Lett. **101**, 013903 (2008)

Chapter 2
Light Propagation in Nonlinear Periodic Media

This chapter describes the fundamentals of light propagation in nonlinear periodic media and provides the basis for all further results presented in this thesis. Starting from general considerations of nonlinear light-matter interaction, the photo-refractive nonlinearity as an ideal mechanism for the combination of periodic refractive index structures with a strong nonlinear material response at moderate laser powers is discussed in detail. Subsequently, the concept of photonic band gaps as a result of periodicity is introduced and the influence of the band structure on linear and nonlinear light propagation is presented.

2.1 Basic Equations of Wave Propagation in Nonlinear Optical Media

We start our discussion by considering Maxwell's equations:

$$\nabla \cdot \boldsymbol{D} = \rho \tag{2.1}$$

$$\nabla \cdot \boldsymbol{B} = 0 \tag{2.2}$$

$$\nabla \times \boldsymbol{E} = -\frac{\partial \boldsymbol{B}}{\partial t} \tag{2.3}$$

$$\nabla \times \boldsymbol{H} = \frac{\partial \boldsymbol{D}}{\partial t} + \boldsymbol{j}. \tag{2.4}$$

In a nonmagnetic material, the electric field $\boldsymbol{E}$ and the magnetic field $\boldsymbol{H}$ are related to the electric displacement field $\boldsymbol{D}$ and the magnetic induction $\boldsymbol{B}$ by the following material equations:

$$\boldsymbol{D} = \epsilon_0 \boldsymbol{E} + \boldsymbol{P} \tag{2.5}$$

B. Terhalle, *Controlling Light in Optically Induced Photonic Lattices*, Springer Theses, DOI: 10.1007/978-3-642-16647-1_2, © Springer-Verlag Berlin Heidelberg 2011

$$\boldsymbol{B} = \mu_0 \boldsymbol{H} \tag{2.6}$$

where ϵ_0 denotes the free space permitivity and $\boldsymbol{P}$ the induced polarization of the material. Since we are interested in solutions to these equations in regions containing no free charges or currents, we set $\rho = 0$ and $\boldsymbol{j} = 0$. Introducing the speed of light in vacuum $c = 1/\sqrt{\mu_0 \epsilon_0}$ and combining (2.3)–(2.6) then leads to the general wave equation

$$\nabla \times \nabla \times \boldsymbol{E} + \frac{1}{c^2}\frac{\partial^2 \boldsymbol{E}}{\partial t^2} = -\frac{1}{\epsilon_0 c^2}\frac{\partial^2 \boldsymbol{P}}{\partial t^2}. \tag{2.7}$$

In this equation, the material's response to the electric field associated with an incident light wave is determined by the polarization $\boldsymbol{P}$. For linear dielectric media, it can be written as

$$\boldsymbol{P} = \epsilon_0 \chi^{(1)} \boldsymbol{E} \tag{2.8}$$

where $\chi^{(1)}$ is the (linear) electric susceptibility of the medium. It should be noted that although it can often be treated as a scalar constant, in the general case of propagation in anisotropic media, $\chi^{(1)}$ is a tensor of rank 2. Substituting (2.8) into (2.5) gives

$$\boldsymbol{D} = \epsilon_0 (1 + \chi^{(1)}) \boldsymbol{E} = \epsilon_0 \epsilon \boldsymbol{E} \tag{2.9}$$

with the dielectric permittivity ϵ. If the electric field of the incident light wave becomes comparable to the intra-atomic field, the linear relation (2.8) is no longer valid. In this case, the polarization is often written as

$$\boldsymbol{P} = \epsilon_0 \left(\chi^{(1)} \boldsymbol{E} + \chi^{(2)} \boldsymbol{E}\boldsymbol{E} + \chi^{(3)} \boldsymbol{E}\boldsymbol{E}\boldsymbol{E} + \cdots \right) \tag{2.10}$$

where $\chi^{(i)}$ with $i \geq 2$ represents the n-th order nonlinear susceptibility. The nonlinear terms in (2.10) give rise to a number of interesting phenomena [1]. For instance, the second-order term is responsible for second harmonic generation or sum- and difference frequency generation while the third-order term causes phenomena like third harmonic generation or optical phase conjugation.

For further analysis, it is convenient to rewrite (2.10) as

$$\boldsymbol{P} = \epsilon_0 \chi_{\mathrm{eff}} \boldsymbol{E} \tag{2.11}$$

with an effective, intensity-dependent susceptibility $\chi_{\mathrm{eff}}(I \equiv |E|^2)$. Using this relation and introducing the effective refractive index $n(I) = \sqrt{1 + \chi_{\mathrm{eff}}}$, the Helmholtz equation can be obtained from (2.7):

$$-\nabla^2 \boldsymbol{E} + \frac{n^2}{c^2}\frac{\partial^2 \boldsymbol{E}}{\partial t^2} = 0. \tag{2.12}$$

In this derivation, we have used the relation $\nabla \times \nabla \times \boldsymbol{E} = \nabla(\nabla \cdot \boldsymbol{E}) - \nabla^2 \boldsymbol{E}$ together with the fact that the term $\nabla(\nabla \cdot \boldsymbol{E})$ is small for most cases of interest [1]. As a next step, the refractive index is divided into its value in the absence of light n_0^2 and the light induced refractive index shift $\Delta n^2(I)$:

$$n^2 = n_0^2 + \Delta n^2(I). \tag{2.13}$$

Moreover, we consider a wave that is linearly polarized in x-direction and propagating in z-direction, i.e.

$$\boldsymbol{E}(\boldsymbol{r}, t) = A(\boldsymbol{r})e^{i(k_z z - \omega t)} \cdot \boldsymbol{e}_x \tag{2.14}$$

with $\boldsymbol{r} = (x, y, z)$ and $k_z = n_o \omega / c$. In the standard paraxial approximation, the envelope $A(x, y, z)$ is assumed to vary with z on a scale much longer than the wavelength. In this case, its second derivative with respect to z can be neglected such that inserting (2.14) into (2.12) results in the paraxial wave equation

$$i\frac{\partial A}{\partial z} + \frac{1}{2k_z}\nabla_\perp^2 A + \frac{k_z}{2n_0^2}\Delta n^2(I)A = 0 \tag{2.15}$$

with $\nabla_\perp^2 = \left(\partial^2/\partial x^2 + \partial^2/\partial y^2\right)$. Introducing the dimensionless variables $x' = x/x_0$, $y' = y/y_0$ and $z' = z/k_z x_0^2$ with a transverse scaling factor x_0 gives

$$2i\frac{\partial A}{\partial z'} + \nabla_\perp'^2 A + \frac{k_z^2 x_0^2}{n_0^2}\Delta n^2(I)A = 0 \tag{2.16}$$

with $\nabla_\perp'^2 = \left(\partial^2/\partial x'^2 + \partial^2/\partial y'^2\right)$. Throughout the rest of this work, we will use this dimensionless form unless otherwise noted. Therefore, we will drop the primes for convenience.

2.2 The Photorefractive Nonlinearity

The photorefractive effect describes the change in the local refractive index of a medium as a result of an optically induced redistribution of charge carriers. It was first observed in 1966 as wavefront distortions of a coherent light beam propagating through a Lithium Niobate (LiNbO$_3$) crystal [2]. Since then, it has been observed in many different materials such as Barium Titanate (BaTiO$_3$), Lithium Tantalate (LiTaO$_3$), Potassium Niobate (KNbO$_3$), and Strontium Barium Niobate (SBN) which will be the material of choice in all experiments presented in this thesis.

The origin of the photorefractive effect is illustrated schematically in Fig. 2.1. Typically, the crystals are doped with acceptors as well as donors. Illuminating them with light of appropriate wavelength lifts up electrons or holes from donors or traps into the conduction band via photoionization. The charges in the conduction band then move under the influence of diffusion, drift in an externally applied field E_0 or the photovoltaic effect and finally recombine with empty donors or traps. This recombination process builts up a static space charge field E_{sc} that modulates the refractive index via the linear electrooptic effect which will be described in the next section.

2.2.1 The Linear Electrooptic Effect

In electrooptic crystals, the presence of a dc electric field leads to a change in the dielectric permittivity ϵ or equivalently a change in dimension and direction of the optical indicatrix [3]. In photorefractive materials, this change depends linearly on the strength of the induced space charge field and is therefore described by the linear electrooptic effect (Pockels effect). The mathematical description of this effect is traditionally given as a change in the impermeability tensor

$$\Delta\eta_{ij} = \Delta\left(\frac{1}{n^2}\right) = \sum_k r_{ijk}E_k.$$ (2.17)

In this notation the constants r_{ijk} are the elements of the electrooptic tensor and E_k represents the k-th component of the total electric field $E = E_0 + E_{sc}$ as the sum of the externally applied field E_0 and the space charge field E_{sc} generated inside the crystal. The corresponding change in the dielectric permittivity is given by

$$\Delta\epsilon_{ij} = -\epsilon_0 n_i^2 n_j^2 \Delta\eta_{ij}.$$ (2.18)

To abbreviate notations, the coefficients r_{ijk} are typically written in a contracted notation defined by the following scheme for $k = 1, 2, 3$:

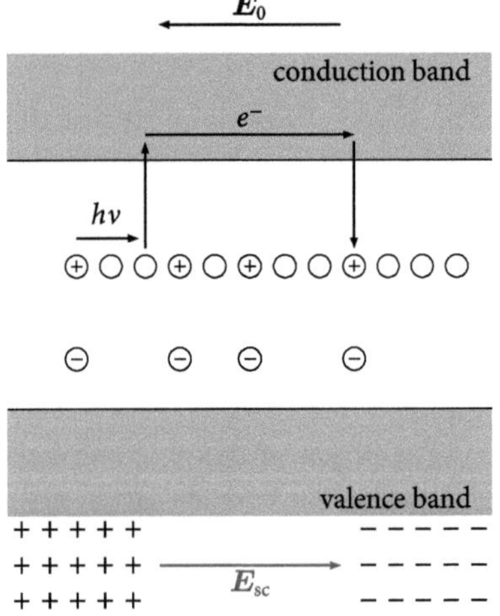

Fig. 2.1 Schematic illustration of charge carrier transport in photorefractive crystals

$$r_{1k} = r_{11k} \qquad\qquad r_{2k} = r_{22k} \qquad\qquad r_{3k} = r_{33k}$$
$$r_{4k} = r_{23k} = r_{32k} \quad r_{5k} = r_{13k} = r_{31k} \quad r_{6k} = r_{12k} = r_{21k}.$$

Throughout this thesis, all experiments will be carried out in Strontium Barium Niobate (SBN) which belongs to the point group 4 mm. In contracted notation, the electrooptic coefficients for this particular symmetry class are given by[1]

$$r_{\text{SBN}} = \begin{bmatrix} 0 & 0 & r_{13} \\ 0 & 0 & r_{13} \\ 0 & 0 & r_{33} \\ 0 & r_{42} & 0 \\ r_{42} & 0 & 0 \\ 0 & 0 & 0 \end{bmatrix}. \tag{2.19}$$

Typically, SBN-crystals obey the relation $r_{33} \gg r_{13}, r_{42}$ such that the light induced changes of the dielectric permittivity ϵ can be described by an effective electrooptic coefficient $r_{\text{eff}} = r_{33} = r_{333}$. If the incident light is extraordinarily polarized, the effective refractive index change is thus determined by

$$\Delta n^2 = -n_0^4 r_{\text{eff}} \boldsymbol{E} \cdot \boldsymbol{e}_c \tag{2.20}$$

where $\boldsymbol{e}_c$ is the unit vector along the direction of the c-axis.

2.2.2 The Band Transport Model

According to (2.20), the light induced refractive index change depends on the effective electro-optic coefficient r_{eff} and the electric field $\boldsymbol{E}$. Therefore, it is crucial to know how the space charge field generated by the incident light looks like. This can be achieved by a simple band transport model which has been developed by Kukhtarev et al. in 1979 to describe the dynamics of the charge carriers inside the crystal [4].

In the following derivation, a photorefractive material of electron-conductivity type is considered and the density of holes is neglected. Similar equations can be derived for materials of the hole-conducting type quite easily, then neglecting the density of electrons.

---

[1] It is assumed that the c-axis of the crystal is oriented along the z-direction of the chosen coordinate system and the space charge field is parallel to this direction. This is the typical choice for the description of electro optic effects. Therefore, it has been adopted here although it is not the common situation throughout the rest of thesis, where propagation along the z-direction is considered. In this case, the crystal is rotated by 90° and the c-axis points in the x-direction.

First, the effective generation rate of electrons is given by the rate of generation of ionized donors minus the rate of recapture of electrons

$$G = (s\tilde{I} + \beta)(N_D - N_D^+) - \alpha_R N_D^+ n_e = \beta(\tilde{I}/I_{sat} + 1)(N_D - N_D^+) - \alpha_R N_D^+ n_e \quad (2.21)$$

where s is proportional to the probability that an incoming photon lifts up an electron into the conduction band (photo excitation) and β describes the excitation of electrons in the absence of light (thermal excitation). N_D, N_D^+ and n_e are the density of donors, ionized donors and free electrons, respectively. Finally, α_R represents the probability for an electron in the conduction band to recombine with an ionized donor.

The saturation intensity I_{sat} determines the ratio between the thermal and the photoionization coefficient which is expressed by a characteristic intensity. This notation is useful because the nonlinear response of a photorefractive material is not only determined by the optical beam intensity $\tilde{I}$, but also by an intensity-equivalent from thermal excitation. Moreover, an artificial contribution to the thermal excitation created by an incoherent background illumination is usually used in experiments to control the behavior of the photorefractive crystal.

The electrons in the conduction band are influenced by three different transport processes: drift in an electric field, diffusion, and the photovoltaic effect. Hence, the resulting current density is given by

$$\begin{aligned} \boldsymbol{j} &= \boldsymbol{j}_{drift} + \boldsymbol{j}_{diff} + \boldsymbol{j}_{pv} \\ &= e\mu n_e \boldsymbol{E} + eD\nabla n_e + \beta_{ph}(N_D - N_D^+)\boldsymbol{e}_c \tilde{I}. \end{aligned} \quad (2.22)$$

Here, $D = \mu k_B T/e$ is the diffusion constant containing the Boltzmann constant k_B, the absolute temperature T, the elementary charge e and the electron mobility μ. The strength of the photovoltaic effect is given by the photovoltaic tensor β_{ph} which has its largest component along the direction of the c-axis, whereas all the other components can be neglected. Finally, $\boldsymbol{e}_c$ gives the unit vector along the c-axis.

Assuming that all acceptors N_A are ionized, the space charge density can be written as

$$\rho = e(N_D^+ - N_A - n_e) \quad (2.23)$$

and Gauß's law [cf. (2.1) and (2.9)] for the present situation becomes

$$\nabla(\epsilon_0 \epsilon \boldsymbol{E}) = \rho = e(N_D^+ - N_A - n_e). \quad (2.24)$$

Since only electrons are considered here as mobile charge carriers and the ionized donor or acceptor atoms are assumed to be stationary within the crystal lattice, (2.23) transforms into the following continuity equation:

$$\frac{\partial \rho}{\partial t} = e\left(\frac{\partial N_D^+}{\partial t} - \frac{\partial n_e}{\partial t}\right) = -\nabla \boldsymbol{j}. \quad (2.25)$$

Furthermore, with this assumption, the change of the density of ionized donors is given by the generation of electrons

$$\frac{\partial N_{\mathrm{D}}^{+}}{\partial t} = G = \beta(I + 1)(N_{\mathrm{D}} - N_{\mathrm{D}}^{+}) - \alpha_{\mathrm{R}} N_{\mathrm{D}}^{+} n_{\mathrm{e}} \tag{2.26}$$

where I denotes the normalized intensity $I = \tilde{I}/I_{\mathrm{sat}}$.

The equations (2.22)–(2.26) form the fundamental set of equations of the Kukhtarev model. To get a simplified expression for the induced space charge field, several approximations that are well justified in most experimental situations can be made. First, a quasi-homogeneous illumination of the crystal is assumed and the system is regarded to be in a steady state ($\partial_t = 0$). Furthermore, the photovoltaic term in (2.22) is neglected.

Based on these assumptions, two major approaches exist which will be described in the following sections. The first one was introduced by Christodoulides and Carvalho [5] and is denoted as *isotropic approximation*. The second one was proposed by Zozulya and Anderson [6] and is referred to as *anisotropic model*.

2.2.3 Isotropic Approximation

In photorefractive materials, the density of impurity ions is typically much larger than the electron density, i.e.

$$N_{\mathrm{D}}^{+}, N_{\mathrm{A}} \gg n_{\mathrm{e}}. \tag{2.27}$$

Using this relation, the density of ionized donor impurities and the free electron density can be derived from (2.24) and (2.26), respectively:

$$N_{\mathrm{D}}^{+} = N_{\mathrm{A}}\left(1 + \frac{\epsilon\epsilon_0}{eN_{\mathrm{A}}}\nabla E\right) \tag{2.28}$$

$$n_{\mathrm{e}} = \frac{\beta(N_{\mathrm{D}} - N_{\mathrm{D}}^{+})}{\alpha_{\mathrm{R}} N_{\mathrm{D}}^{+}}(1 + I). \tag{2.29}$$

Assuming that the intensity I asymptotically attains a constant value I_∞ at the crystal border and the space charge field in this region is also independent of the transverse coordinates, i.e. $|E(x \to \pm\infty)| = |E_0|$, the free electron density $n_{\mathrm{e}}^{(0)}$ at the border can be derived from (2.29):

$$n_{\mathrm{e}}^{(0)} = \frac{\beta(N_{\mathrm{D}} - N_{\mathrm{A}})}{\alpha_{\mathrm{R}} N_{\mathrm{A}}}(1 + I_\infty). \tag{2.30}$$

Since the system is assumed to be in a steady state, the current density (2.22) has to be constant everywhere, i.e. $n_{\mathrm{e}}^{(0)} E_0 = n_{\mathrm{e}} E + D\mu^{-1}\nabla n_{\mathrm{e}}$ or equivalently

$$E = \frac{n_{\mathrm{e}}^{(0)} E_0}{n_{\mathrm{e}}} - \frac{D}{\mu n_{\mathrm{e}}}\nabla n_{\mathrm{e}}. \tag{2.31}$$

Considering drift-dominated charge carrier migration, the diffusion term can be neglected. Further on, assuming that the beam intensity varies slowly with respect to the spatial coordinates $(\nabla E_{sc} \approx 0)$ and setting $I_\infty = 0$, the combination of (2.29), (2.30) and (2.31) results in

$$E = E_0 \frac{1}{1+I}. \tag{2.32}$$

It is striking that the electric field in (2.32) does not depend on any spatial coordinate. This is why the model is denoted as *isotropic approximation*. It is quite appropriate for all one-dimensional configurations and the study of saturable nonlinear media in general, but it does not take into account the specific anisotropic properties of biased photorefractive media. To get a more realistic model of the physical process, it is therefore necessary to use the anisotropic model which will be described in the next section.

2.2.4 Anisotropic Model

Applying an external electric field in one transverse dimension inherently causes a symmetry breaking and destroys the isotropic situation. As a consequence, the isotropic approximation is no longer applicable and a model which includes the distinct anisotropic properties of biased photorefractive crystals has to be used. Such a model was first introduced by Zozulya and Anderson [6] in 1994.

The basic idea is to express the electric field in terms of its electrostatic potential:

$$E = -\nabla \tilde{\phi}. \tag{2.33}$$

Considering an external field E_0 applied across the transverse x-direction, the potential $\tilde{\phi}$ consists of a light induced term ϕ and an external bias term $-|E_0|x$, thus $\tilde{\phi} = \phi - |E_0|x$. As before, it is assumed that the intensity varies slowly with respect to the spatial coordinates. With this assumption, (2.29) transforms into

$$n_e = \frac{\beta(N_D - N_A)}{z'N_A}(1+I). \tag{2.34}$$

Inserting (2.34) into (2.22) and neglecting the photovoltaic term for a steady state system $(\nabla \mathbf{j} = 0)$ then yields

$$\nabla^2 \phi + \nabla \ln(1+I)\nabla\phi = |E_0|\frac{\partial \ln(1+I)}{\partial x}$$

$$- D\left[\nabla^2 \ln(1+I) + (\nabla \ln(1+I))^2\right]. \tag{2.35}$$

For the case of two transverse dimensions, this equation has no analytical solutions and thus has to be solved numerically.[2] The resulting potential can then be used to obtain the total electric field $E = E_0 - \nabla\phi$ and the refractive index change according to (2.20). However, when solving the propagation equation (2.16), the refractive index change $\Delta n^2(I)$ is often determined directly from the light-induced space charge field $E_{sc} = -\nabla\phi$. This can be done because the contribution of E_0 only gives an additional phase factor and neglecting this term is therefore equivalent to rescaling n_0.

In the transversally one-dimensional limit (all y-derivatives equal to zero), an analytical solution of (2.35) can be found:

$$E_{sc} = \left(-|E_0| \frac{I}{1+I} - \frac{D}{1+I}\frac{\partial I}{\partial x} \right) \cdot e_x. \tag{2.36}$$

In fact, this one-dimensional solution is often taken as an approximation in the two-dimensional case. Neglecting the diffusion term $(D = 0)$ then results in the same expression for the total electric field $E = E_{sc} + E_0$ as obtained from the isotropic approximation presented in the previous section [cf. (2.32)]. Such an isotropic approach can obviously not describe the anisotropic features of the photorefractive nonlinearity which have already been observed in several experiments [7–9]. In the following, we will therefore use the anisotropic model (2.35) and again neglect the diffusion term such that

$$\nabla^2\phi + \nabla \ln(1+I)\nabla\phi = |E_0|\frac{\partial \ln(1+I)}{\partial x}. \tag{2.37}$$

The propagation equation (2.16) is finally written as

$$2i\frac{\partial A}{\partial z} + \nabla_{\perp}^2 A - \Gamma E_{sc}(|A|^2)A = 0 \tag{2.38}$$

where we have introduced the coupling constant $\Gamma = k_z^2 x_0^2 n_0^2 r_{eff}$. The space charge field E_{sc} is treated as a scalar quantity, since only the component parallel to the c-axis contributes to the refractive index change.

2.3 Periodic Photonic Structures

In a periodic photonic structure, the refractive index n is periodically modulated along one or more spatial dimensions. The simplest examples of such structures are one-dimensional Bragg gratings which have been known for more than 100 years now and are widely used as filters that reflect optical waves incident at

---

[2] A suitable algorithm is described in appendix A.

certain angles or waves at certain frequencies [10]. Following the pioneering works of Yablonovitch [11] and John [12], dielectric structures with a spatially modulated refractive index in one, two or even three dimensions are nowadays commonly denoted as photonic crystals. This notation is due to the existence of photonic band gaps (see below), representing an optical analog of electronic band gaps in crystalline solids (e.g. semiconductors).

Throughout the rest of this work, we consider transversally periodic structures being homogeneous in the direction of propagation and the coordinate system is chosen such that the light propagates in z-direction (Fig. 2.2). Typical examples for the one-dimensional case include dielectric mirrors and arrays of optical waveguides [13], whereas two-dimensional structures are commonly represented by photonic crystal fibers [14, 15] and planar photonic crystals [16] as shown in Fig. 2.2c, d.

Currently, there exists a number of different approaches for the fabrication of photonic crystals in one or more spatial dimensions, e.g. focused ion beam milling, e-beam lithography combined with reactive ion etching, or two-photon polymerization. However, all these techniques are resource demanding and cost ineffective and thus not well suited for our fundamental investigations relying on flexible modification of structural parameters. Furthermore, the access to nonlinear effects is often hampered by high power requirements. In Chap. 3, we will demonstrate how optically induced photonic lattices can be employed to overcome these limitations, providing highly reconfigurable photonic structures with a strong nonlinear response at very low power levels.

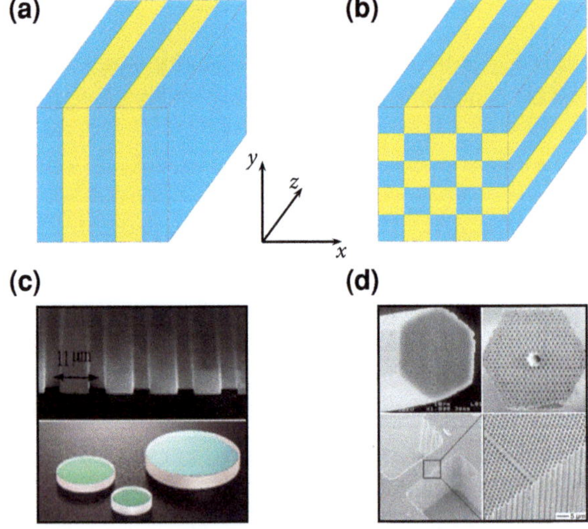

Fig. 2.2 Schematic illustration (*top*) and real fabricated examples (*bottom*) of transversally periodic photonic structures being homogeneous in propagation direction. (**a**), (**c**) One-dimensional case; (**b**), (**d**) two-dimensional case; pictures in (**c**) and (**d**) taken from [14–17]

2.3.1 Band Gap Spectrum

To introduce the basic concepts of wave propagation in transversally periodic structures, we first assume a static, prefabricated refractive index modulation. The special case of optically induced photonic lattices will be discussed in Chap. 3.

Including a transversally periodic, static refractive index modulation $\Delta n_{\mathrm{pot}}^2$ results in an additive potential term in the propagation equation (2.38) which then reads as

$$2i\frac{\partial A}{\partial z} + \nabla_{\perp}^2 A + V(x,y)A - \Gamma E_{\mathrm{sc}}(|A|^2)A = 0 \tag{2.39}$$

with $V(x,y) = k_z^2 x_0^2 n_0^{-2} \Delta n_{\mathrm{pot}}^2$. It is important to note that this equation shows strong formal analogies to the description of matter waves in optical lattices. The reason for this is that by choosing $E_{\mathrm{sc}}(|A|^2) = \pm|A|^2$ (Kerr-type nonlinearity) and substituting $z \to t$, (2.39) transforms into a two-dimensional form of the Gross-Pitaevskii equation describing a Bose-Einstein condensate in a two-dimensional optical lattice [18].

Furthermore, the linearized version of (2.39) (i.e. $\Gamma = 0$) is the standard equation for the description of electrons in periodic electronic potentials in solid state physics.

Inserting the ansatz $A(\boldsymbol{r}) = \tilde{A}(\boldsymbol{r}_{\perp}, \boldsymbol{k}_{\perp}) \cdot e^{i\beta z}$ with $\tilde{A}(\boldsymbol{r}_{\perp}, \boldsymbol{k}_{\perp}) = a(\boldsymbol{r}_{\perp}, \boldsymbol{k}_{\perp}) \cdot e^{i\boldsymbol{k}_{\perp} \cdot \boldsymbol{r}_{\perp}}$, this linearized equation transforms into the eigenvalue problem

$$\frac{1}{2}\Big[(\nabla_{\perp} + i\boldsymbol{k}_{\perp})^2 + V(\boldsymbol{r}_{\perp})\Big]a(\boldsymbol{r}_{\perp}, \boldsymbol{k}_{\perp}) = \beta(\boldsymbol{k}_{\perp})a(\boldsymbol{r}_{\perp}, \boldsymbol{k}_{\perp}) \tag{2.40}$$

with $\boldsymbol{r}_{\perp} = (x,y)$ and $\boldsymbol{k}_{\perp} = (k_x, k_y)$. β is the propagation constant which can be seen as an offset to the wave vector component k_z [cf. (2.14)].

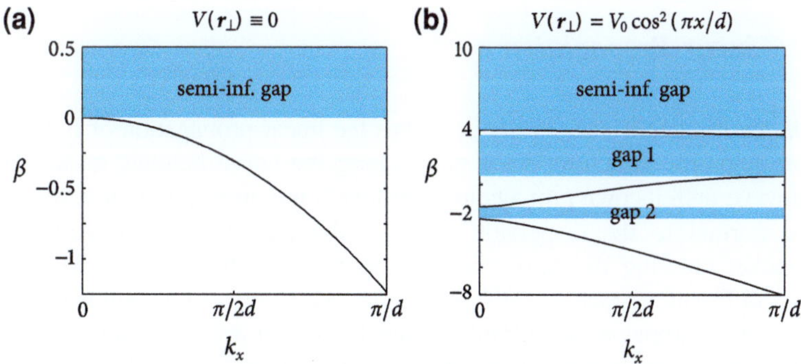

Fig. 2.3 Calculated linear dispersion relations. **a** Bulk homogeneous medium with $V(\boldsymbol{r}_\perp) \equiv 0$; **b** one-dimensional lattice with $V(\boldsymbol{r}_\perp) = V_0 \cdot \cos^2(\pi x/d)$, $V_0 = 0.0009$, $d = 2x_0$

In homogeneous bulk media $(V(\boldsymbol{r}_\perp) \equiv 0)$, the eigenmode solutions of the system are plane waves $\tilde{A}(\boldsymbol{r}) = a_0 \cdot e^{i\boldsymbol{k}_\perp \cdot \boldsymbol{r}_\perp}$ with $a_0 = $ const. and we obtain the dispersion relation $\beta(\boldsymbol{k}_\perp) = -(k_x^2 + k_y^2)/2$. This parabolic relation is plotted in Fig. 2.3a. It shows a semi-infinite gap of forbidden propagation constants which may be seen as an optical analog of the forbidden energy regime below the ground state of an electron in a solid state. Note that the absolute value of β is only determined up to an arbitrary additional constant, since the total wave vector component in z-direction is given by $k_z + \beta$ and there is no fixed rule for the choice of k_z in the initial ansatz (2.14). Throughout this thesis, we adopt a common choice in which the semi-infinite gap reaches from 0 to $+\infty$ for $V_0 \to 0$.

In the presence of a periodic modulation $V(\boldsymbol{r}_\perp)$, the dispersion gets strongly modified due to multiple Bragg-reflections within the structure. As a result, the dispersion curve $\beta(\boldsymbol{k}_\perp)$ is divided into several transmission bands separated by forbidden gaps of finite size. The eigenmode solutions to the linearized version of (2.39) are in this case given by the well-known Bloch waves

$$\tilde{A}_m(\boldsymbol{r}_\perp, \boldsymbol{k}_\perp) = a_m(\boldsymbol{r}_\perp, \boldsymbol{k}_\perp) \cdot e^{i\boldsymbol{k}_\perp \cdot \boldsymbol{r}_\perp} \tag{2.41}$$

where the functions $a_m(\boldsymbol{r}_\perp, \boldsymbol{k}_\perp)$ have the periodicity of the underlying lattice and the index $m = 1, 2, \ldots$ denotes the band number. If we are not referring to a specific band, we will drop this index for convenience in the following. Figure 2.3b shows the resulting dispersion relation for a one-dimensional modulation $V(x, y) = V_0 \cdot \cos^2(\pi x / d)$ with effective modulation depth V_0 and lattice constant d. Note that in general the dispersion relation shows the translation invariance $k_{x,y} \to k_{x,y} \pm 2\pi/d$. Therefore, it is fully defined by its values in the first Brillouin zone $k_{x,y} \in [-\pi/d, \pi/d]$. Moreover, the first Brillouin zone itself is often redundant since periodic structures often possess additional symmetries. By eliminating these redundant regions, one obtains the so-called irreducible Brillouin zone. For the one-dimensional structures represented in Fig. 2.3, this irreducible Brillouin zone is given by $k_x \in [0, \pi/d]$.

2.3.2 Linear Propagation

The dispersion curve $\beta(\boldsymbol{k}_\perp)$ fully describes the linear propagation of light through a given photonic structure. Each point along the curve belongs to an extended Bloch wave with its own propagation constant β and propagation direction defined by the normal to the dispersion curve at the particular point. Any wave or wavepacket entering the lattice is decomposed into these Bloch waves which acquire different phases as they propagate. Such an accumulation of different phases during propagation certainly affects the waveform and results in an output that may considerably differ from the input waveform.

As an example, we consider the diffraction of a spatially finite beam covering a certain spectral range $[\Delta k_x, \Delta \beta]$. The diffractive properties of such a wavepacket are governed by the relative spread or convergence of adjacent waves which can be determined from the curvature of the dispersion curve. In regions of convex curvature, the beam acquires a convex phasefront, resulting in normal diffraction with wave behavior similar to that in homogeneous media. In contrast, exciting a group of waves in a region of concave band curvature leads to anomalous diffraction, i.e. the input beam acquires a concave wavefront during propagation. For the dispersion relation shown in Fig. 2.3b, this means that beams centered at the top of the first band and propagating along the lattice will experience normal diffraction ($\partial^2 \beta / \partial k_x^2 < 0$) while beams corresponding to the bottom of the first band will experience anomalous diffraction ($\partial^2 \beta / \partial k_x^2 > 0$). Between the two regions of normal and anomalous diffraction, a region of zero curvature exists where $\partial^2 \beta / \partial k_x^2 = 0$. Waves belonging to this zone will therefore experience no linear diffraction broadening. As a result, the diffraction properties can be controlled by changing the incident angle. This feature has been denoted as *diffraction management* in analogy to dispersion management of optical pulses in fibers [19].

In addition, the effect of the lattice on the wave propagation also depends on the transverse size of the input beam relative to the lattice spacing. For example, a broad Gaussian beam launched at zero angle into the lattice excites modes from different bands and stays mostly Gaussian as it propagates. In contrast, coupling a narrow beam into the fundamental guided mode of a single period of the lattice excites Bloch modes primarily from the first band. In this case, due to coupling between lattice sites and multiple interference effects, the beam undergoes *discrete diffraction* characterized by intense side lobes with little or no light in the central region [20, 21]. The corresponding intensity distribution is shown in Fig. 2.4a for propagation in an AlGaAs waveguide array [20].

2.3.3 Nonlinear Propagation

As demonstrated in Sect. 2.1, the refractive index and thus the wave vector of light propagating in a nonlinear medium depends on the intensity. Therefore, in a nonlinear periodic structure, the wave vector component in propagation direction can be shifted by the nonlinearity such that, compared to the linear regime, it can

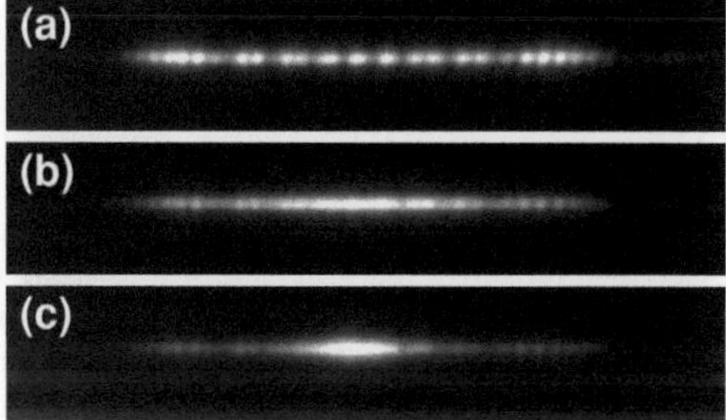

Fig. 2.4 First experimental observation of a discrete soliton in an AlGaAs waveguide array (reprinted from [20]). **a** Discrete diffraction at low power; **b** partial self-trapping at intermediate power; **c** formation of a discrete soliton at high power

be located in a gap and nonlinear propagation inside the linear gaps becomes possible. In particular, these nonlinear gap waves can be localized and form a *discrete or gap soliton* if diffraction is compensated by nonlinearity. In 1998, Eisenberg et al. [20] realized the first experimental observation of optical discrete solitons by focusing femtosecond laser pulses into a one-dimensional AlGaAs waveguide array. While at low peak power, the already discussed discrete diffraction pattern was observed (Fig. 2.4a), increasing the peak powers led to a narrowing of the transverse intensity distribution and finally the formation of a discrete soliton as shown in Fig. 2.4b, c, respectively.

Similar to the case of linear propagation, the influence of nonlinearity again depends on the curvature of the dispersion relation. In regions of normal diffraction, the resulting convex curvature of the phase front can be compensated by a focusing nonlinearity ($E_{sc} > 0$) while regions of anomalous diffraction require a defocusing nonlinearity ($E_{sc} < 0$). Consequently, such a dependence enables manipulation of the nonlinear dynamics by engineering the band structure in much the same way as linear diffraction management [22].

Gap solitons can be considered as self-guided modes of nonlinearity induced positive or negative defects in the periodic structure (Fig. 2.5a, b). Such defects naturally have localized modes (rather than the extended Bloch modes of the ideal lattice), whose propagation constant now lies off the linear transmission band, i.e. in a gap. For a focusing nonlinearity, the locally increased refractive index and thus the increased propagation constant causes a positive defect as sketched in Fig. 2.5a. The negative defect resulting from a defocusing nonlinearity is illustrated in Fig. 2.5b.

Depending on the gap, the self guiding property of the gap soliton modes has two different origins. On the one hand, it can arise from the locally increased refractive index in the presence of a focusing nonlinearity, similar to homogeneous

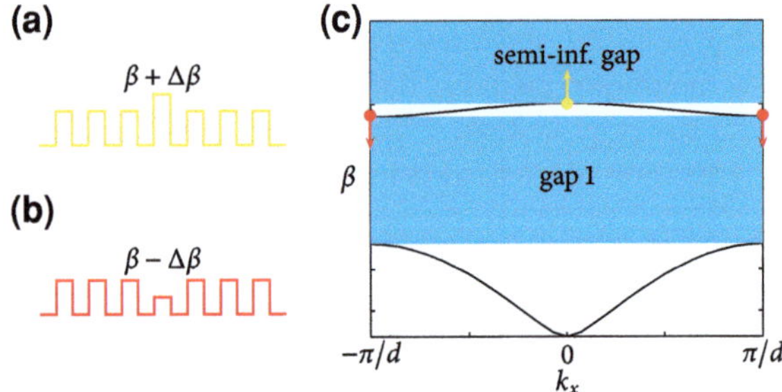

Fig. 2.5 Schematic illustration of discrete and gap soliton formation originating from the first band in a one-dimensional periodic photonic structure. **a** Positive defect caused by the locally increased refractive index in case of a focusing nonlinearity; **b** negative defect for a defocusing nonlinearity; **c** calculated dispersion relation with arrows indicating the nonlinear shift of the propagation constant: (*yellow*) positive shift, formation of a discrete soliton; (*red*) negative shift, formation of a gap soliton in the presence of a defocusing nonlinearity

media. This is the case for solitons in the semi-infinite gap which are commonly named *discrete solitons*. In this case, the propagation constant originating from the top of the first band is shifted up into the gap (yellow arrow in Fig. 2.5c). On the other hand, multiple transverse Bragg reflections may also enable self-guided modes. This is the dominant mechanism in all finite gaps and the corresponding solitons are often denoted as *gap solitons*. Unfortunately, the use of these terms is not consistent in literature. However, throughout the rest of this thesis, we will keep the most common choice of notation and refer to the solitons in the semi-infinite gap as discrete solitons while using the term gap solitons for localization in all finite gaps.

To give an example of a gap soliton, we consider Bloch waves from the first band at the edge of the Brillouin zone of a one-dimensional periodic structure at $k_x = \pm \pi / d$. The corresponding soliton has a propagation constant lying between the first and the second band [23]. For modes originating from the first band, a defocusing nonlinearity is necessary for the soliton formation and the propagation constant goes down into the gap with increasing nonlinearity as indicated by the red arrows in Fig. 2.5c.

In the context of this thesis, the term soliton describes a spatial amplitude profile which does not change during propagation, i.e.

$$A(\boldsymbol{r}) = \psi(\boldsymbol{r}_\perp) \cdot e^{i\beta z}. \tag{2.42}$$

Substituting this expression into the propagation equation (2.39) results in the following equation for the soliton profile $\psi(\boldsymbol{r}_\perp)$ in a static, prefabricated photonic structure:

$$-2\beta\psi + \nabla_\perp^2 \psi + V(x,y)\psi - \Gamma E_{sc}(|\psi|^2)\psi = 0. \tag{2.43}$$

The corresponding equation for the case of optically induced photonic lattices will be given in Chap. 3. Due to the complex structure of the photorefractive nonlinearity, analytical solutions to (2.43) do not exist and thus it has to be solved numerically. A suitable algorithm to achieve this is described in [24].

References

1. Boyd, R.W.: Nonlinear Optics, 3rd edn. Academic press, Elsevier (2008)
2. Ashkin, A., Boyd, G.D., Dziedzic, J.M., Smith, R.G., Ballmann, A.A., Levin- stein, J.J., Nassau, K.: Optically-induced refractive index inhomogeneities in $LiNbO_3$ and $LiTaO_3$. Appl. Phys. Lett. **9**, 72 (1966)
3. Yeh, P.: Introduction to Photorefractive Nonlinear Optics. Wiley, New York (1993)
4. Kukhtarev, N.V., Markov, V.B., Odoulov, S.G., Soskin, M.S., Vinetskii, V.L.: Holo-graphic storage in electro-optic crystals: I. steady state + II. beam coupling and light amplification. Ferroelectrics **22**, 961 (1979)
5. Christodoulides, D.N., Carvalho, M.I.: Bright, dark, and gray spatial soliton states in photorefractive media. J. Opt. Soc. Am. B **12**, 1628 (1995)

6. Zozulya, A.A., Anderson, D.Z.: Propagation of an optical beam in a photorefractive medium in the presence of a photogalvanic nonlinearity or an externally applied electric field. Phys. Rev. A **51**, 1520 (1995)
7. Desyatnikov, A.S., Sagemerten, N., Fischer, R., Terhalle, B., Träger, D., Neshev, D.N., Dreischuh, A., Denz, C., Krolikowski, W., Kivshar, Y.S.: Two-dimensional self-trapped nonlinear photonic lattices. Opt. Express **14**, 2851 (2006)
8. Terhalle, B., Desyatnikov, A.S., Bersch, C., Träger, D., Tang, L., Imbrock, J., Kivshar, Y.S., Denz, C.: Anisotropic photonic lattices and discrete solitons in photorefractive media. Appl. Phys. B **86**, 399 (2007)
9. Krolikowski, W., Saffman, M., Luther-Davies, B., Denz, C.: Anomalous interaction of spatial solitons in photorefractive media. Phys. Rev. Lett. **80**, 3240 (1998)
10. Yeh, P.: Optical Waves in Layered Media. Wiley, New York (1998)
11. Yablonovitch, E.: Inhibited spontaneous emission in solid-state-physics and electronics. Phys. Rev. Lett. **58**, 2059 (1987)
12. John, S.: Strong localization of photons in certain disordered dielectric superlattices. Phys. Rev. Lett. **58**, 2486 (1987)
13. Sukhorukov, A.A., Kivshar, Y.S., Eisenberg, H.S., Silberberg, Y.: Spatial optical solitons in waveguide arrays. IEEE J. Quantum Electron. **39**, 31 (2003)
14. Birks, T.A., Knight, J.C., Russell, P.S.J.: Endlessly single-mode photonic crystal fiber. Opt. Lett. **22**, 961 (1997)
15. Cregan, R.F., Mangan, B.J., Knight, J.C., Birks, T.A., Russell, P.S.J., Roberts, P.J., Allan, D.C.: Single-mode photonic band gap guidance of light in air. Science **285**, 1537 (1999)
16. Birner, A., Wehrspohn, R.B., Gösele, U.M., Busch, K.: Silicon-based photonic crystals. Adv. Mater. **13**, 377 (2001)
17. Neshev, D.N., Sukhorukov, A.A., Krolikowski, W., Kivshar, Y.S.: Nonlinear optics and light localization in periodic photonic lattices. J. Nonlinear Opt. Phys. Mater. **16**, 1 (2007)
18. Ostrovskaya, E.A., Kivshar, Y.S.: Matter-wave gap solitons in atomic band-gap structures. Phys. Rev. Lett. **90**, 160407 (2003)
19. Eisenberg, H.S., Silberberg, Y., Morandotti, R., Aitchison, J.S.: Diffraction management. Phys. Rev. Lett. **85**, 1863 (2000)
20. Eisenberg, H.S., Silberberg, Y., Morandotti, R., Boyd, A.R., Aitchison, J.S.: Discrete spatial optical solitons in waveguide arrays. Phys. Rev. Lett. **81**, 3383 (1998)
21. Fleischer, J., Bartal, G., Cohen, O., Schwartz, T., Manela, O., Freedman, B., Segev, M., Buljan, H., Efremidis, N.: Spatial photonics in nonlinear waveguide arrays. Opt. Express **13**, 1780 (2005)
22. Ablowitz, M.J., Musslimani, Z.H.: Discrete diffraction managed spatial solitons. Phys. Rev. Lett. **87**, 254102 (2001)
23. Kivshar, Y.S.: Self-localization in arrays of defocusing waveguides. Opt. Lett. **18**, 1147 (1993)
24. Richter, T.: Stability of anisotropic gap solitons in photorefractive media. PhD thesis, TU Darmstadt (2008)

Chapter 3
Optically Induced Photonic Lattices

In this chapter, we provide a detailed description of optically induced structures in photorefractive media as the experimental platform used throughout this thesis. Based on the concept of nondiffracting periodic waves, different experimental approaches for the realization of two-dimensional photonic structures as well as the analysis of the induced lattices are discussed. It is demonstrated that the photorefractive anisotropy significantly influences the symmetry of the resulting refractive index change. Finally, the important case of hexagonal structures is discussed in more detail and it is shown that in this case the influence of anisotropy can be compensated by stretching the lattice wave along its vertical direction.

3.1 The Optical Induction Technique

As already pointed out before, the optical induction in photorefractive media provides an ideal tool to create highly nonlinear, reconfigurable photonic lattices for fundamental investigations of wave propagation in periodic structures [1].

The basic idea is to modify the refractive index of a bulk nonlinear material by external illumination. If such an illumination is spatially periodic, as in interference patterns, the material will acquire a periodic modulation of the refractive index, thus resembling a photonic crystal. An additional probe beam can be employed to study wave propagation within the resulting periodic structure. The probe beam should be incoherent to the periodically shaped lattice wave and propagate in the same direction. Thus, it experiences a periodic refractive index modulation and the response of the medium is still nonlinear.

The problem with this approach is that in general the lattice wave will be affected by the refractive index modulation itself and also experience nonlinear self-action at sufficiently high intensities. As a result, the lattice structure will inevitably vary along the propagation direction in nonlinear media of significant thickness. Furthermore, in the presence of the copropagating probe beam, strong

B. Terhalle, *Controlling Light in Optically Induced Photonic Lattices*, Springer Theses, DOI: 10.1007/978-3-642-16647-1_3, © Springer-Verlag Berlin Heidelberg 2011

interaction via cross-phase modulation (XPM) will occur [2]. Although it has been demonstrated that in this configuration stable lattices are possible within certain parameter regions [3, 4], and cross-phase modulation may even facilitate localized composite states [5], such nonlinear self-action of the lattice beam as well as nonlinear interaction with the probe is often not desirable.

In 2002, Efremidis et al. resolved this problem by realizing that the photore-fractive nonlinearity in certain materials, such as SBN, shows a strong *polarization anisotropy*. This means that the strength of the nonlinearity "seen" by a light wave depends on its polarization or equivalently the corresponding electrooptic coeffi-cient r_{ij} [cf. (2.20)]. Because of $r_{13} \ll r_{33}$, an ordinarily polarized lattice wave in SBN propagates in an effectively linear regime and does not show any self-action. However, it still induces the desired periodic refractive index modulation, because the excitation of charge carriers into the conduction band only depends on the intensity and is not affected by polarization [cf. (2.21)]. At low intensities, the induced refractive index change then acts as a linear periodic potential on the probe beam which remains extraordinarily polarized. However, if the intensity of the probe beam is sufficiently increased, it also modifies the refractive index and propagates nonlinearly through the periodic structure. Hence, depending on the intensity of the prope beam, linear as well as nonlinear propagation effects can be studied in this system.

The modified propagation equation (2.38) for the case of optically induced photonic lattices reads as

$$2i\frac{\partial A}{\partial z} + \nabla_\perp^2 A - \Gamma E_{\text{sc}}(I_{\text{tot}})A = 0 \tag{3.1}$$

with the total Intensity $I_{\text{tot}} = |A_{\text{latt}}|^2 + |A|^2$ determined by the lattice intensity $|A_{\text{latt}}|^2$ and the probe beam intensity $|A|^2$. The dispersion relation $\beta(k_x, k_y)$ for this case can be obtained by making the same ansatz as in Sect. 2.3 using $I_{\text{tot}} = |A_{\text{latt}}|^2$. Discrete and gap solitons are again found by inserting (2.42) into the propagation equation (3.1). In this case, the resulting equation for the soliton profile $\psi(\mathbf{r}_\perp)$ in an optically induced photonic lattice is given by

$$-2\beta\psi + \nabla_\perp^2 \psi - \Gamma E_{sc}(|A_{\text{latt}}|^2 + |\psi|^2)\psi = 0. \tag{3.2}$$

Similar to the case of prefabricated lattices, no analytical solution exists and (3.2) has to be solved numerically [6].

In order to achieve the desired two-dimensional lattice structures (cf. Fig. 2.2), the transversally periodic intensity distribution has to be stationary along the propagation direction. For the interference of plane waves, this implies that each plane wave is represented in the transverse Fourier space (k_x, k_y) by a point located on a ring centered at the origin $k_x = k_y = 0$. The reason for this is schematically illustrated in Fig. 3.1. The interference of two plane waves with wave vectors $\mathbf{k}_1$ and $\mathbf{k}_2$ results in a modulated intensity distribution along $\mathbf{K} = \mathbf{k}_1 - \mathbf{k}_2$. This intensity distribution will be stationary along the propagation direction for $K_z = 0$.

Therefore, the plane waves represented by the wave vectors k_i must have the same longitudinal wave vector component k_{iz}. For a fixed wavelength λ, the relation

$$|k_i| = \sqrt{k_{ix} + k_{iy} + k_{iz}} = \frac{2\pi}{\lambda} \tag{3.3}$$

implies that wave vectors with the same longitudinal component k_{iz} are located on a ring in the transverse Fourier space. The same argumentation holds for the general case of more than two plane waves. Interference of any combination of plane waves fulfilling (3.3) will result in a stationary, transversally modulated lattice wave suitable for the optical induction of two-dimensional photonic structures.

3.1.1 Experimental Realizations

A typical setup for the realization of nondiffracting periodic lattice waves for the optical induction of two-dimensional photonic structures is shown schematically in Fig. 3.2.

The beam from a frequency-doubled Nd:YAG laser at a wavelength of 532 nm is sent through a combination of half wave plate and polarizer to control the intensity and ensure ordinary polarization. It is then split into four beams by passing two subsequent Mach–Zehnder-type interferometers. Depending on the desired lattice structure, individual beams may be blocked to interfere either two, three or all four beams inside a 20 mm long photorefractive $Sr_{0.60}Ba_{0.40}Nb_2O_6$ (SBN:Ce) crystal which is externally biased with a dc electric field directed along its c-axis. Two CCD cameras are used to analyze the light exiting the crystal in real space as well as in Fourier space. The period and the spatial orientation of the periodic intensity distribution can be controlled by adjusting the relative angles of the interfering beams.

Obviously, the number of available transverse lattice geometries is limited by the maximum number of interfering waves. For example, the above setup enables

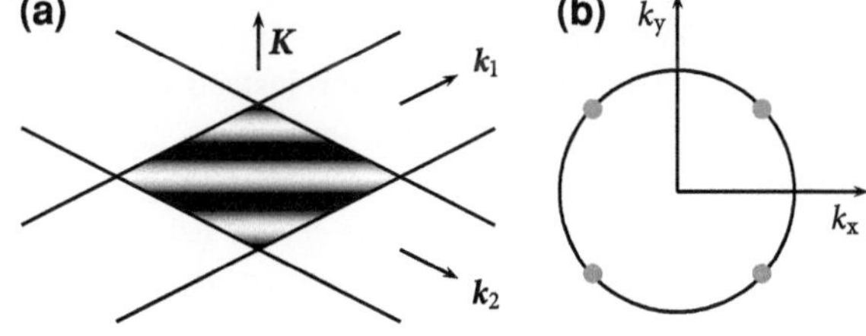

Fig. 3.1 Schematic representation of plane wave interference for optical induction of photonic structures. **a** Two plane waves k_1 and k_2 forming a modulated intensity distribution with grating vector $K = k_1 - k_2$; **b** four plane waves having the same longitudinal wave vector component k_z

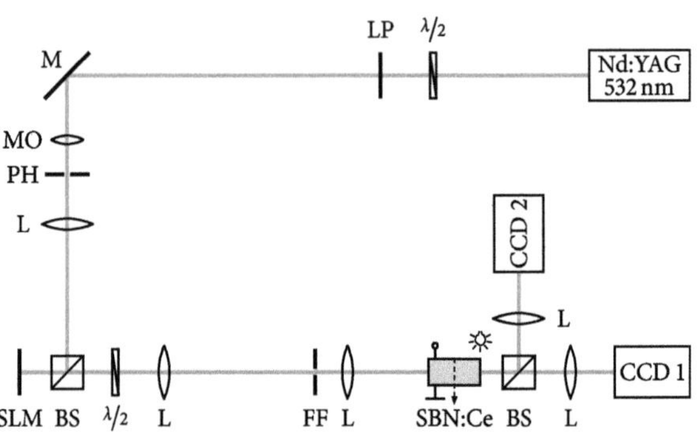

Fig. 3.2 Experimental setup for optical induction of transversally periodic photonic structures inside a photorefractive material using plane wave interference. *BS* beam splitter; *CCD* camera; *L* lens; *LP* linear polarizer; *M* mirror; *MO* microscope objective; *PH* pinhole

Fig. 3.3 Experimental setup for optical induction of transversally periodic photonic structures inside a photorefractive material using a spatial light modulator. *BS* beam splitter; *CCD* camera; *FF* Fourier filter; *L* lens; *LP* linear polarizer; *M* mirror; *MO* microscope objective; *PH* pinhole; *SLM* spatial light modulator

the induction of one-dimensional stripe patterns by interfering two waves as well as two-dimensional lattices of hexagonal or square symmetry by using three or four waves, respectively. In principle, the number of beams can of course be increased by additional Mach–Zehnder configurations to achieve more complex lattice geometries. However, such additional beams significantly hamper the experimental handling and structures requiring more than four interfering waves are usually not realized this way.

A very powerful way to overcome these limitations is to use a programmable spatial light modulator instead of the interferometer configurations. A sketch of the corresponding experimental setup is shown in Fig. 3.3. As before, the laser beam is sent through the combination of half wave plate and polarizer, but now illuminates a spatial light modulator instead of passing through the interferometer configurations. The modulator imprints a phase modulation corresponding to a certain lattice geometry onto the incoming wave and the modulated beam is imaged onto the front face of the crystal using a demagnifying telescope.[1] A mask in the Fourier plane of the telescope ensures that higher order Fourier components

---

[1] An experimental procedure to determine the phase characteristics of the modulator is described in Sect. 7.2.

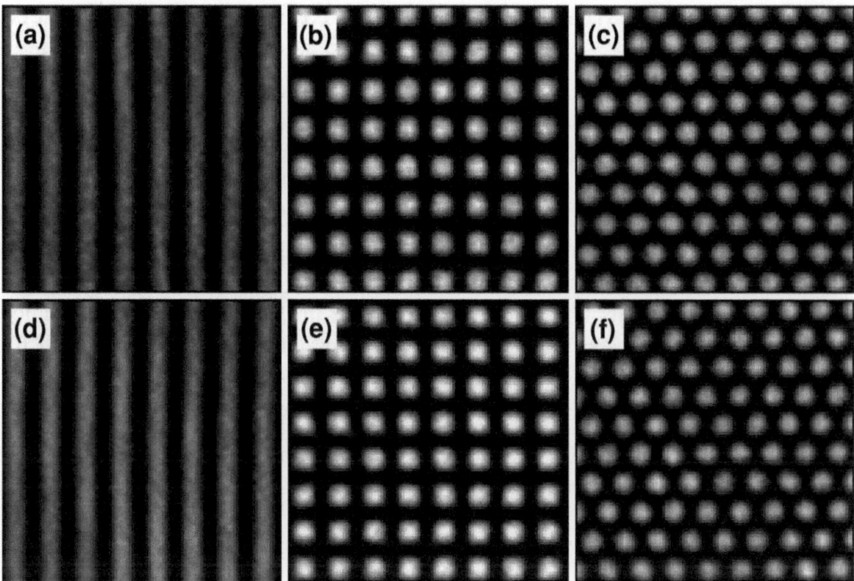

Fig. 3.4 Transversally periodic intensity distribution at the front face (*top*) and back face (*bottom*) of the crystal. **a, d** Stripe pattern; **b, e** square pattern; **c, f** hexagonal pattern

are filtered out such that the resulting amplitude and phase structure of the lattice wave is equivalent to the corresponding interference of plane waves. The output of the crystal is again analyzed in Fourier space as well as in real space using two CCD cameras.

The main advantage of this method is given by its high flexibility. In contrast to the previous setup, a change of parameters like lattice period and orientation or switching between different lattice geometries does not require any adjustment of optical components but can easily be achieved by modifying the imprinted phase modulation and the corresponding Fourier filtering. Even complex lattice geometries consisting of advanced non-diffracting waves [7] or multiperiodic structures as a superposition of several singleperiodic waves [8] can be realized. However, studying the properties of such complex lattices is not the aim of this thesis.

In general, both presented methods enable the induction of all the structures investigated throughout this work. Nevertheless, due to its high flexibility, the use of a spatial light modulator is still preferable. Unless otherwise noted, we will therefore use the setup shown in Fig. 3.3 but, depending on the experimental situation, the interference setup depicted in Fig. 3.2 may also be used.

As an example, Fig. 3.4 illustrates the experimentally obtained lattice waves for a one-dimensional stripe pattern as well as two-dimensional patterns of square and hexagonal geometry. To demonstrate the stationary, transversally periodic intensity distribution, images are taken at the front face as well as the back face of the crystal for each lattice geometry.

3.2 Structure Analysis of the Induced Refractive Index Patterns

In the last section, it has been shown how to create a nondiffracting periodic lattice wave for optical induction of photonic lattices in photorefractive media. To get a deeper understanding of the induced lattices, it is necessary to develop methods for the analysis of the refractive index change inside the crystal.

Numerical simulations based on (2.37) and (2.38) provide a complete description of the beam propagation inside the photorefractive crystal and hence allow the calculation of the induced refractive index change for arbitrary propagation distances z. This is generally not possible in the experimental realization, where only front face and back face of the crystal can be imaged onto the CCD camera. Due to this restriction, the experimental diagnostic tools have to be able to provide information about the induced refractive index change inside the crystal by analyzing the intensity distributions at front and back face, respectively.

3.2.1 Waveguiding

The simplest way to analyze the induced refractive index structure is to exploit its waveguiding properties [3, 4]. If the lattice is illuminated with a broad plane wave, this wave is guided in the regions of high refractive index and the output intensity pattern of the guided beam qualitatively maps the induced refractive index change. More precisely, this method corresponds to the excitation of the Bloch wave belonging to the first band, having an amplitude modulated with the same symmetry as the underlying refractive index lattice. Since SBN crystals have a relatively slow response time at low intensities, the output of the plane wave can quickly be monitored without modifying the induced refractive index change. In particular, the illumination of the induced lattice with a plane wave can be achieved with the setup shown in Fig. 3.3 by simply switching off the modulator. This results in an unmodulated plane wave which is then sent onto the front face of the crystal by the telescope. The two CCD cameras finally enable a direct analysis of the induced refractive index change in real space as well as in Fourier space.

3.2.2 Brillouin Zone Spectroscopy

Another important method to investigate the structure of the induced refractive index change in Fourier space is given by Brillouin zone spectroscopy [9, 10].

As explained in Sect. 2.3, a periodic modulation of the refractive index modifies the diffraction relation and leads to regions of allowed propagation separated by forbidden band gaps which open up at the boundaries of the Brillouin zones. This behavior is generic for wave propagation in periodic structures and not restricted to optical systems. However, in optical systems, the appearance of band gaps in the transmission spectrum at the borders of the Brillouin zones enables a direct visualization of the lattice structure in Fourier space since the regions of forbidden propagation appear as dark lines in the power spectrum of the transmitted beam [10]. To achieve this, the induced lattice has to be probed with a broad spectrum of the two-dimensional Boch waves. Additionally, the probe beam has to be broad enough in real space to cover a sufficient number of elementary cells. These requirements can be fulfilled simultaneously by a partially spatially incoherent probe beam, which enables a homogeneous excitation of several Brillouin zones with a beam that occupies numerous lattice sites [10].

Such a partially spatially incoherent beam is created experimentally by passing a laser beam through a diffuser which is rotating on a time scale much faster than the response time of the photorefractive nonlinearity. The diffuser deforms the initially homogeneous phase front in an irregular way and leads to a sum of random contributions from various parts of the diffuser at any point. When the diffuser is rotated, the optical field changes randomly with time resulting in a fluctuating intensity comparable to thermal light. If it rotates at sufficiently high speed, the camera and the crystal only detect a time-averaged intensity and the power spectrum becomes smooth, because the speckles are washed out. As a result, the induced photonic lattice is probed with a partially spatially incoherent beam which has a uniform spatial power spectrum extending over several Brillouin zones.

To perform Brillouin zone spectroscopy experimentally, the setup used for optical induction and waveguiding (Fig. 3.3) is modified as shown in Fig. 3.5. The polarizer is replaced by a polarizing beam splitter to get two beams with adjustable intensities. The transmitted beam is used for the optical induction of photonic lattices in the same way as before. The reflected beam is passed through the rotating diffuser and the partially spatially incoherent output of the diffuser is imaged onto the front face of the crystal. Its polarization can be adjusted by a half wave plate and its power spectrum is controlled using an iris diaphragm in the common focal plane of the two imaging lenses. Finally, the output of the crystal is analyzed in Fourier space using CCD2. The real space camera CCD1 is not used in this case. In spite of this, it is drawn here because this way Fig. 3.5 shows a

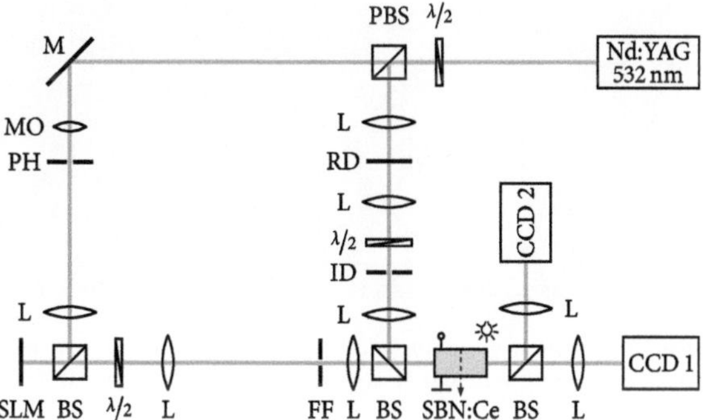

Fig. 3.5 Experimental setup for Brillouin zone spectroscopy of optically induced photonic lattices. *CCD* camera; *FF* Fourier filter; *ID* iris diaphragm; *L* lens; *M* mirror; *MO* microscope objective; *(P)BS* (polarizing) beam splitter; *PH* pinhole; *RD* rotating diffuser; *SLM* spatial light modulator

complete setup for structure analysis of the induced refractive index change which enables observation of the waveguiding properties of the induced lattice as well as Brillouin zone spectroscopy.

3.2.3 Orientation Anisotropy

In the following, the experimental methods mentioned above will be used to investigate the distinct anisotropic properties of two-dimensional optically induced photonic lattices. The experimental results will be accompanied by numerical simulations using the anisotropic model. As a first example, we consider a lattice wave of square symmetry as it has already been shown in Fig. 3.4b and e. In this case, the lattice wave can be written as $A_{\text{latt}}(x, y) = \sin(k_x x) \sin(k_y y)$ with the spatial frequencies $k_x = k_y = 2\pi/d$ and the lattice constant d. Using this expression, the refractive index change Δn^2 is obtained numerically from (2.20) and (2.37). Surprisingly, as a result of the photorefractive anisotropy, the induced lattice shows an effectively one-dimensional structure, although the original lattice wave is fully two-dimensional (Fig. 3.6a). In experiment, the lattice is induced by using a lattice wave at a power of $P_{\text{latt}} = 37\,\mu\text{W}$ and applying an electric field of $E_0 = 1.4\,\text{kV/cm}$ across the crystal. The lattice constant is $d = 16\,\mu\text{m}$. The obtained

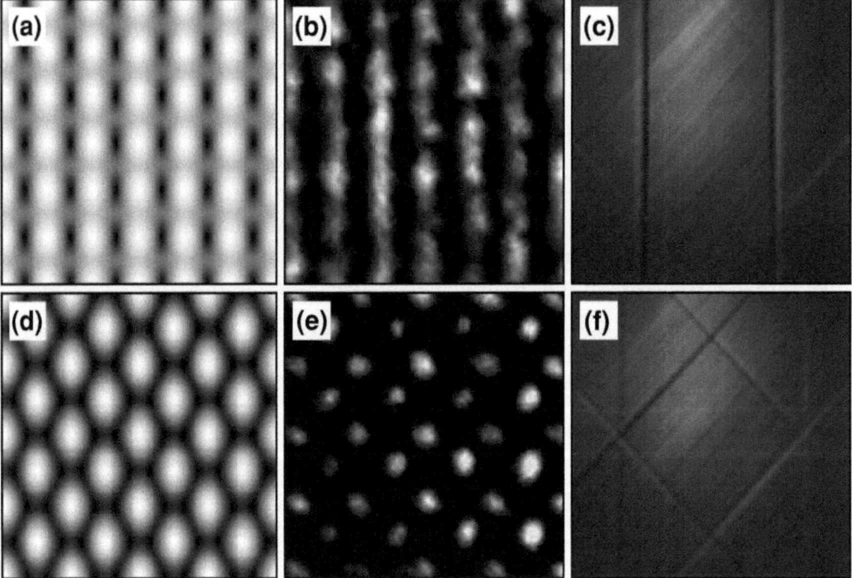

Fig. 3.6 Structure analysis of the square lattice (*top*) and the diamond lattice (*bottom*). **a, d** Numerically calculated refractive index change; **b, e** waveguiding; **c, f** Brillouin zone spectroscopy

results of waveguiding and Brillouin zone spectroscopy shown in Fig. 3.6b and c clearly confirm the one-dimensional refractive index change predicted by the numerical simulations.

The influence of anisotropy can be minimized if the lattice wave is rotated by 45°; a configuration which is also known as *diamond pattern* in literature. In this case, the lattice wave is given by $A_{\text{latt}}(x, y) = \sin\left((k_x x + k_y y)/2\right) \cdot \sin\left((k_x x - k_y y)/2\right)$ and the numerical simulations reveal the two-dimensional refractive index pattern shown in Fig. 3.6d. For the experimental analysis, we use the same parameters as before, except for the rotation of the lattice wave. In contrast to the square pattern (Fig. 3.6b), the waveguiding output now shows a two-dimensional waveguide structure (Fig. 3.6e) and the Brillouin zone picture in Fig. 3.6f confirms this result.

As demonstrated by this simple example of a square lattice, the induced refractive index change is not only determined by the geometry of the lattice wave but also depends strongly on its spatial orientation and this may cause significant symmetry reductions. Similar effects of *orientation anisotropy* can be observed for other lattice types [3, 7].

For the rest of the thesis, our experiments will mainly focus on lattices with hexagonal symmetry. Therefore, a detailed description of this lattice type including its anisotropic features will be given in the next section.

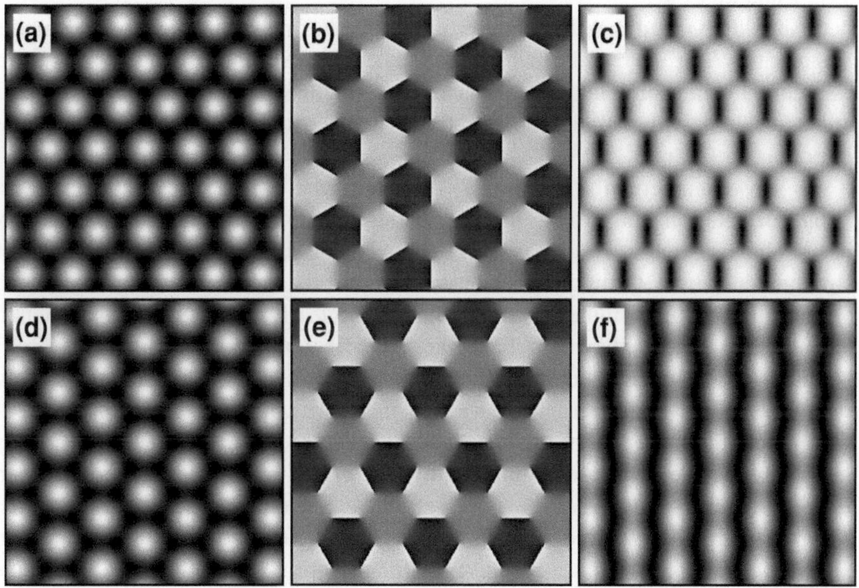

Fig. 3.7 Numerically calculated lattice wave and refractive index change of the hexagonal lattice in horizontal (*top*) and perpendicular orientation (*bottom*). **a, d** Intensity distribution; **b, e** phase structure; **c, f** refractive index change

3.3 Hexagonal Lattices

We focus on hexagonal lattices for several reasons: first, among the possible two-dimensional geometries, the hexagonal one has been shown to be advantageous by supporting larger band gaps [11]. Hence, most prefabricated planar photonic crystal structures possess hexagonal symmetry. Moreover, the stacking method fabrication of photonic crystal fibers naturally results in the same geometry [12]. From a more general point of view, interest in the hexagonal lattice arises from the fact that it is not separable. This means it cannot be represented as a sum of two one-dimensional lattices.

Similar to the square and diamond pattern, two orientations with respect to the c-axis have to be distinguished for the hexagonal lattice as well. In the parallel orientation (Fig. 3.7a), the lattice spots are aligned along the horizontal direction (parallel to the c-axis), while in the perpendicular one, they are aligned along the vertical direction (Fig. 3.7d).

According to the experimental situation, the lattice wave for the horizontal orientation can be written in the general form of three interfering plane waves

$$A_{\text{latt}} = A_0 \left[e^{(2ik_x x/3)} + e^{\left(-ik_x x/3 + ik_y y\right)} + e^{\left(-ik_x x/3 - ik_y y\right)} \right] \tag{3.4}$$

with equal amplitudes A_0 and the transverse wave vector components $k_x = \sqrt{3}k_y$. From this expression, the perpendicular orientation is obtained by substituting $x \leftrightarrow y$ and $k_x \leftrightarrow k_y$.

It should be noted that the phase profile of the hexagonal lattice waves contains singularities (Fig. 3.7b and e) which are located between the lattice sites and form a so-called honeycomb lattice [3]. Such phase singularities or optical vortices [13] are associated with circular energy flows in the transverse plane and will be important in Chap. 5.

As shown numerically in Fig. 3.7, the hexagonal lattice is also affected by a strong orientation anisotropy. Similar to the square pattern (Fig. 3.8), the perpendicular orientation results in an effectively one-dimensional refractive index structure in which the individual lattice sites fuse together forming modulated vertical lines. This symmetry reduction is also clearly seen in the experimental waveguiding output shown in Fig. 3.8e. The Brillouin zone picture (Fig. 3.8f) confirms this result by showing two dominant vertical lines. To achieve a truly two-dimensional lattice structure, we will therefore choose the horizontal orientation throughout the rest of this work. In this case, the fusing into vertical lines is prevented as demonstrated numerically in Fig. 3.7c and experimentally in Fig. 3.8b and c. However, the photorefractive anisotropy still causes a much stronger modulation between two spots along the c-axis (horizontal direction) than along the "diagonals". As a result, even for the horizontal orientation, the symmetry of the lattice wave is not preserved completely and the induced refractive index change shows a rhombus-like structure in which the coupling between the lattice sites is much larger along the diagonals than along the horizontal direction.

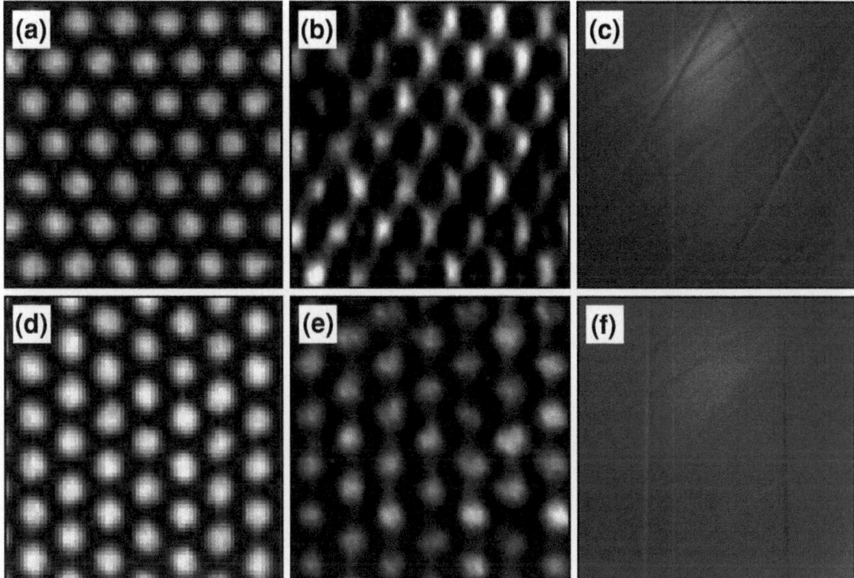

Fig. 3.8 Structure analysis of the hexagonal lattice (lattice constant $d = 14\,\mu$m) in horizontal (*top*) and perpendicular orientation (*bottom*). **a, d** Lattice wave; **b, e** waveguiding; **c, f** Brillouin zone spectroscopy; experimental parameters: $P_{\text{latt}} = 30\,\mu$W, $E_0 = 1.4\,$kV/cm

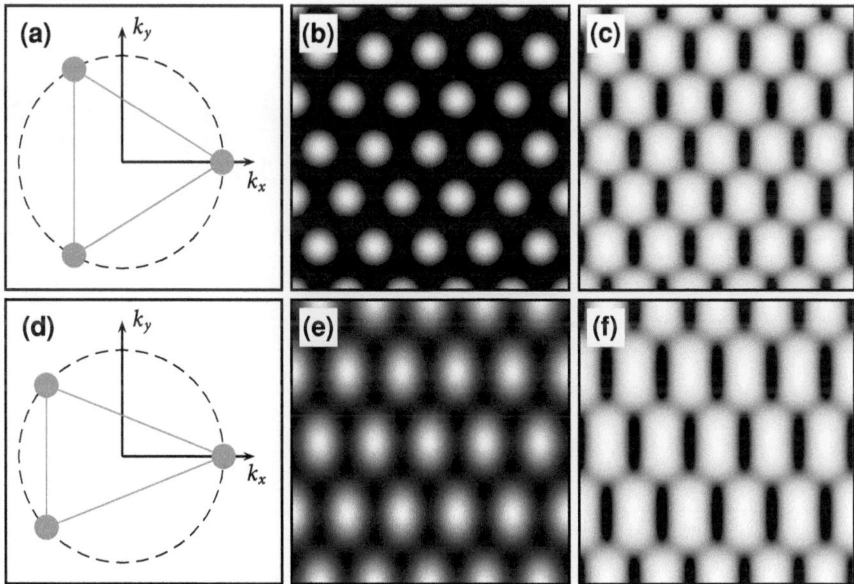

Fig. 3.9 Sketch of the Fourier image and numerically calculated lattice intensity and refractive index profiles for the hexagonal lattice. **a–c** Symmetric lattice; **d–f** stretched lattice

The rhombus-like structure is clearly demonstrated by the experimental results of waveguiding and Brillouin zone spectroscopy shown in Fig. 3.8b and c, respectively. Since the coupling between the lattice sites decreases with larger spatial separation, this effect can be compensated by stretching the lattice wave along the vertical direction, thus achieving a more symmetric coupling closer to that of the original hexagonal symmetry. More specifically, we introduce the stretching factor $\eta = k_x/k_y = d_y/d_x$ where d_x and d_y denote the lattice constants along the x and y direction, respectively. Fig. 3.9 shows the lattice wave and the refractive index change for the symmetric hexagonal lattice ($\eta = \sqrt{3}$) and a stretched lattice with $\eta = 2.5$. In the transverse Fourier space (k_x, k_y), the lattice beams form an equilateral triangle for the unstretched lattice (Fig. 3.9a) and an isosceles triangle for the stretched lattice (Fig. 3.9d). In both cases, they are still located on a circle centered at the origin $k_x = k_y = 0$. Thus, the nondiffracting propagation (cf. Sect. 3.1) is preserved. The calculated dispersion curves for both symmetries are shown in Fig. 3.10 together with the corresponding refractive index modulations and sketches of the first Brillouin zones. By exploiting the mirror symmetries, the Brillouin zones can be reduced to the so-called irreducible Brillouin zone marked by the yellow trapezoid. It is important to note that, due to the photorefractive anisotropy, the symmetry points X_1 and X_2 are not equivalent and therefore the Brillouin zone cannot be further reduced. In an isotropic system, this could be done and the irreducible Brillouin zone would be given by the triangle $\Gamma M_1 X_1$.

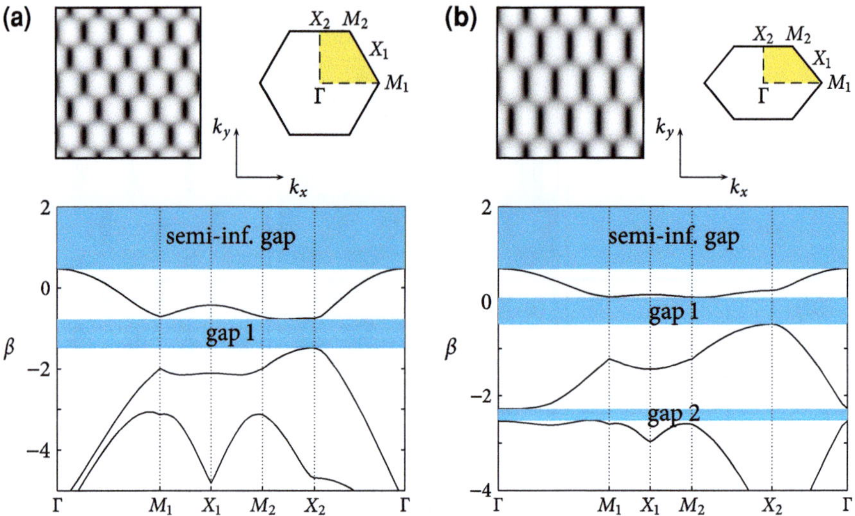

Fig. 3.10 Calculated linear dispersion relations for **a** the symmetric hexagonal lattice with $\eta = \sqrt{3}$ and **b** the stretched lattice with $\eta = 2.5$. For both cases, the calculated refractive index change (*top left*) and a sketch of the Brillouin zone (*top right*) are shown as well. The irreducible Brillouin zone defined by the corresponding high symmetry points is marked in *yellow*

References

1. Efremidis, N.K., Sears S., Christodoulides D.N., Fleischer J.W., Segev, M.: Discrete solitons in photorefractive optically induced photonic lattices. Phys. Rev. E **66**, 046602 (2002)
2. Saleh, B.E.A., Teich, M.C.: Fundamentals of Photonics. Wiley, New York (2007)
3. Desyatnikov, A.S., Sagemerten, N., Fischer, R., Terhalle, B., Träger, D., Neshev, D.N., Dreischuh, A., Denz, C., Krolikowski, W., Kivshar, Y.S.: Two-dimensional self-trapped nonlinear photonic lattices. Opt. Express **14**, 2851–2863 (2006)
4. Desyatnikov, A.S., Neshev, D.N., Kivshar, Y.S., Sagemerten, N., Träger, D., Jägers, J., Denz, C., Kartashov, Y.V.: Nonlinear photonic lattices in anisotropic nonlocal self-focusing media. Opt. Lett. **30**, 869–871 (2005)
5. Desyatnikov, A.S., Ostrovskaya, E.A., Kivshar, Y.S., Denz, C.: Composite band-gap solitons in nonlinear optically induced lattices. Phys. Rev. Lett. **91**, 153902 (2003)
6. Richter, T., Stability of anisotropic gap solitons in photorefractive media. PhD thesis, TU Darmstadt (2008)
7. Rose, P., Richter, T., Terhalle, B., Imbrock, J., Kaiser, F., Denz, C.: Discrete and dipole-mode gap solitons in higher-order nonlinear photonic lattices. Appl. Phys. B **89**(4), 521–526 (2007)
8. Rose, P., Terhalle, B., Imbrock, J., Denz, C.: Optically induced photonic superlattices by holographic multiplexing. J. Phys. D: Appl. Phys. **41**, 224004 (2008)
9. Terhalle, B., Trager, D., Tang, L., Imbrock, J., Denz, C.: Structure analysis of two-dimensional nonlinear self-trapped photonic lattices in anisotropic photorefractive media. Phys. Rev. E **74**, 057601 (2006)
10. Bartal, G., Cohen, O., Buljan, H., Fleischer, J.W., Manela, O., Segev, M.: Brillouin zone spectroscopy of nonlinear photonic lattices. Phys. Rev. Lett. **94**, 163902 (2005)
11. Joannopoulus, J.D., Meade, R.D., Winn, J.N.: Photonic Crystals: Molding the Flow of Light. Princeton University Press, Princeton (1995)
12. Knight, J.C., Birks, T.A., Russell, P.S.J., Atkin, D.M.: All-silica single-mode optical fiber with photonic crystal cladding. Opt. Lett. **21**, 1547–1549 (1996)
13. Desyatnikov, A.S., Torner, L., Kivshar, Y.S.: Optical vortices and vortex solitons. Prog. Optics **47**, 219–319 (2005)

Chapter 4
Resonant Rabi Oscillations and Interband Transitions

In this chapter, we first review the derivation of the generalized Landau-Zener-Majorana (LZM) model as a simple theoretical basis which allows to predict and understand the subsequent experimental observations. We discuss in detail the experimental observations of two- and three-level Rabi oscillations triggered by Bragg resonances as well as the onset of Landau-Zener tunneling when an additional linear refractive index gradient is superimposed onto the two-dimensional hexagonal lattices. Furthermore, we investigate the influence of nonlinearity, facilitating interband coupling and triggering modulational instability. The experimental results are compared to the predictions of the resonant LZM systems as well as anisotropic numerical simulations.

4.1 Introduction

Due to the similarities between light propagation in periodic refractive index structures and electrons in crystalline solids (cf. Sect. 2.3), many effects which were originally predicted in solid state physics have recently been observed in optics [1]. Two fundamental phenomena in this context are *Bloch oscillations* [2] and *Zener tunneling* [3]. Both are associated with the propagation of waves or quantum particles in periodic potentials under the action of an external driving force and were originally predicted for electrons moving in a periodic potential with a superimposed constant electric field. In this situation, the particles do not just follow the driving force but perform an oscillatory motion known as Bloch oscillation. However, these oscillations do not persist forever, but are damped by interband transitions known as Zener tunneling [4]. Examples of this tunneling process include the electrical breakdown in Zener diodes [5] or electrical conduction in semiconductor superlattices [6].

Optical analogs of Bloch oscillations and Zener tunneling have been demonstrated for one-dimensional structures based on waveguide arrays [7–9] or superlattices [10, 11]. In such optical settings, the periodic modulation of the refractive

B. Terhalle, *Controlling Light in Optically Induced Photonic Lattices*, Springer Theses, DOI: 10.1007/978-3-642-16647-1_4, © Springer-Verlag Berlin Heidelberg 2011

index plays the role of the crystalline potential, while an additional linear refractive index ramp acts as external force. Recently, the experimental observations have been extended to two dimensions by the observation of Bloch oscillations and Zener tunneling in optically induced square lattices [12].

The theoretical description of Zener tunneling in one-dimensional periodic structures is given by a model which was independently employed by Landau [13], Zener [3], and Majorana [14] for the study of electrical breakdown and quantum tunneling in 1932. Based on these studies, a generalized Landau-Zener-Majorana (LZM) model has been derived for two-dimensional photonic lattices of square [15] as well as hexagonal symmetry [16, 17].

Using this generalized LZM model, it has also been demonstrated that without the additional linear gradient, wave propagation through a periodic photonic structure may result in power oscillations between the Fourier peaks related by the Bragg condition. Due to formal analogies with the oscillations of level populations in a two-level quantum system under the influence of an incident light field, the observed power oscillations in Fourier space have been denoted as Bragg resonance induced Rabi oscillations [18].

In this chapter, we will start by briefly reviewing the derivation of the generalized LZM model and subsequently present the first experimental observation of the above mentioned Rabi oscillations triggered by Bragg resonances as well as Zener tunneling in two-dimensional hexagonal lattices. The experimental results will be accompanied by anisotropic numerical simulations and both, experiment and numerics will be compared to the theoretical results obtained from the generalized LZM model.

4.2 The Landau-Zener-Majorana Model

We start by considering the linearized propagation equation for a shallow (small amplitude) periodic potential $V(\mathbf{r}_\perp)$ with an additional linear tilt $\boldsymbol{\alpha}\mathbf{r}_\perp$ in the form

$$i\frac{\partial A}{\partial z} = -\frac{1}{2}\nabla_\perp^2 A + [V(\mathbf{r}_\perp) + \boldsymbol{\alpha}\mathbf{r}_\perp]A. \tag{4.1}$$

As already mentioned in the previous section, we focus on two-dimensional lattice potentials of hexagonal symmetry which can be achieved by three interfering plane waves as $V = V_0 \sum_{j=1}^{3} e^{i\mathbf{k}_j\mathbf{r}_\perp}$. Using a more general approach, the expression for a hexagonal structure (not necessarily created by interference) can be given as the following Fourier series:

$$V = V_0\left\{ \varepsilon_1 \sum_{l=1}^{3} \cos(\boldsymbol{b}_l\mathbf{r}_\perp) + \varepsilon_2 \sum_{l=1}^{3} \cos(2\boldsymbol{b}_l\mathbf{r}_\perp). \right.$$

$$\left. +\varepsilon_3\left[\cos[(\boldsymbol{b}_1 + \boldsymbol{b}_2)\mathbf{r}_\perp] + \cos[(\boldsymbol{b}_2 + \boldsymbol{b}_3)\mathbf{r}_\perp] + \cos[(\boldsymbol{b}_3 - \boldsymbol{b}_1)\mathbf{r}_\perp]] + \cdots \right\} \tag{4.2}$$

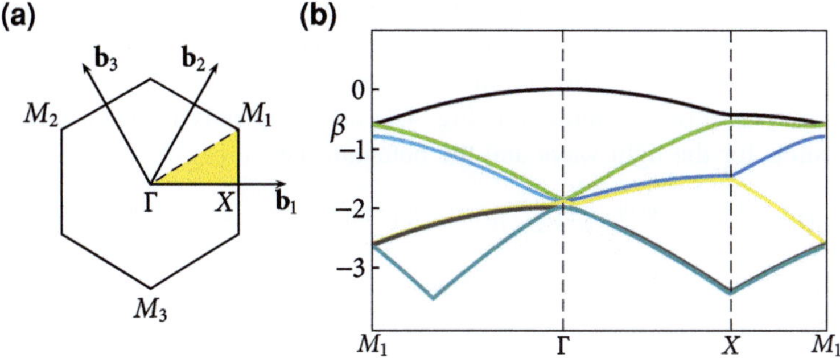

Fig. 4.1   a First Brillouin zone and **b** band structure of the hexagonal lattice (4.3) for $b = |\boldsymbol{b}_j| = 2$ and $V_0 = 0.1$

with the base vectors $\boldsymbol{b}_1 = \boldsymbol{k}_1 - \boldsymbol{k}_2, \boldsymbol{b}_2 = \boldsymbol{k}_1 - \boldsymbol{k}_3$, and $\boldsymbol{b}_3 = \boldsymbol{b}_2 - \boldsymbol{b}_1 = \boldsymbol{k}_2 - \boldsymbol{k}_3$.

Choosing $\epsilon_1 = 3/2, \epsilon_2 = -1/4$, and $\epsilon_3 = 1/2$ while setting all other terms to zero, we obtain the hexagonal lattice as

$$V = 8V_0 \cos^2\left(\frac{\boldsymbol{b}_1 \boldsymbol{r}_\perp}{2}\right) \cos^2\left(\frac{\boldsymbol{b}_2 \boldsymbol{r}_\perp}{2}\right) \cos^2\left(\frac{\boldsymbol{b}_3 \boldsymbol{r}_\perp}{2}\right). \tag{4.3}$$

For $|\boldsymbol{\alpha}| = 0$, the solutions to (4.1) are defined by the Bloch waves $A(\boldsymbol{r}) = \tilde{A}(\boldsymbol{r}_\perp, \boldsymbol{k}_\perp) \cdot e^{i\beta z}$ as described in Sect. 2.3. The corresponding band structure $\beta(\boldsymbol{k}_\perp)$ for the hexagonal lattice (4.3) is plotted in Fig. 4.1. For a small index gradient, $|\boldsymbol{\alpha}| \ll 1$, the bands are preserved and one can study interband transitions analytically using perturbation theory [19] for $|V_0| \ll 1$. In this limit of a shallow lattice, (4.1) is conveniently analyzed directly in Fourier space, rather than using the Bloch wave approach described in Sect. 2.3. Therefore, we follow [16] by setting

$$A(\boldsymbol{r}_\perp, z) = \int \mathrm{d}\boldsymbol{k}_\perp C(\boldsymbol{k}_\perp, z) e^{i(\boldsymbol{k}_\perp - \boldsymbol{\alpha} z)\boldsymbol{r}_\perp} \tag{4.4}$$

and expanding the lattice potential into the Fourier series

$$V(\boldsymbol{r}_\perp) = \sum_{\boldsymbol{Q}} \hat{V}_{\boldsymbol{Q}} e^{i\boldsymbol{Q}\boldsymbol{r}_\perp}. \tag{4.5}$$

Inserting (4.4) and (4.5) into (4.1) then gives

$$i\frac{\mathrm{d}C(\boldsymbol{k}_\perp, z)}{\mathrm{d}z} = \beta_0(\boldsymbol{k}_\perp - \boldsymbol{\alpha} z)C(\boldsymbol{k}_\perp, z) + \sum_{\boldsymbol{Q}} \hat{V}_{\boldsymbol{Q}} C(\boldsymbol{k}_\perp - \boldsymbol{Q}, z) \tag{4.6}$$

where $\beta_0(\boldsymbol{q}) = \frac{1}{2}\boldsymbol{q}^2$ and $\boldsymbol{Q}$ is the reciprocal lattice vector. For further analysis, we write $\boldsymbol{k}_\perp = \boldsymbol{q} - \boldsymbol{Q}$ with $\boldsymbol{q}$ in the first Brillouin zone and note that, due to the lattice, the Fourier coefficients $C(\boldsymbol{q} - \boldsymbol{Q})$ and $C(\boldsymbol{q} - \boldsymbol{Q}')$ are effectively coupled (about some point z_0) with $\beta_0(q - Q - \alpha z_0) \approx \beta_0(q - Q' - \alpha z_0)$. This implies that interband transitions occur at the Bragg resonance point z_0.

We focus our investigation on the experimentally feasible three-fold resonance which couples three M points (cf. Fig. 4.1) and keep only resonant terms in the expressions for the light wave and the potential, i.e.

$$A_{\text{res}}(r) = C_1(z)e^{iq(z)r_\perp} + C_2(z)e^{i[q(z)+q_{M_2}-q_{M_1}]r_\perp} + C_3(z)e^{i[q(z)+q_{M_3}-q_{M_1}]r_\perp} \quad (4.7)$$

$$V_{\text{res}}(r_\perp) = \mathcal{V}_0\left[e^{i(q_{M_2}-q_{M_1})r_\perp} + e^{i(q_{M_3}-q_{M_1})r_\perp} + e^{i(q_{M_3}-q_{M_2})r_\perp} + c.c. \right] \quad (4.8)$$

with $\mathcal{V}_0 = 3V_0/4$ and q_{M_1}, q_{M_2}, and q_{M_3} connecting the Γ and M points [17]. The Fourier amplitudes C_j are related to the three M points by $C_1 \to M_1, C_2 \to M_2$, and $C_3 \to M_3$.

We require cancelation of the linear terms in $r_\perp$ which gives $dq/dz = -\alpha$ and set $C_j = e^{i\phi(z)}c_j$ with $d\phi/dz = -(q_{M_1}^2 + \alpha^2 z^2)/2$. This way, inserting the resonant terms in (4.1) results in the generalized Landau-Zener-Majorana model as the following system of coupled differential equations [16, 17]:

$$i\frac{d}{dz}c = H(\alpha, z)c \quad (4.9)$$

with

$$c = \begin{pmatrix} c_1 \\ c_2 \\ c_3 \end{pmatrix}, \quad H(\alpha, z) = \begin{pmatrix} -v_1 z & \mathcal{V}_0 & \mathcal{V}_0 \\ \mathcal{V}_0 & -v_2 z & \mathcal{V}_0 \\ \mathcal{V}_0 & \mathcal{V}_0 & -v_3 z \end{pmatrix}, \quad v_j = q_{M_j}\alpha. \quad (4.10)$$

Note that the system (4.9) and (4.10) is invariant under $\pi/3$-rotation according to the hexagonal symmetry.

To illustrate the relation between (4.9) and the band structure $\beta(k_\perp)$, we set the Bragg resonance at $z_0 = 0$ and consider the Bloch bands at the M_1 point [cf. Fig. 4.1]. The three lowest bands along the α direction are determined by the adiabatic eigenvalues $\lambda_j(z)$ of $H(\alpha, z)$ as $\beta_j = -(q_{M_1}^2 + \lambda_j(z))$.

With $z_0 = 0$, the eigenvalues are found in the standard way by solving the characteristic equation $\det[H - \lambda(z)\mathcal{I}] = 0$ with the unit matrix $\mathcal{I}$. As a result, the first two bands collide with $\beta_{1,2} = -q_{M_1}^2 + \mathcal{V}_0$ while the third band is given by $\beta_3 = -q_{M_1}^2 - 2\mathcal{V}_0$. For large z, i.e. away from the resonance, the eigenvalues are approximated as $-v_j z$, thus leading to $\beta_j = -q_{M_1}^2 + v_j z$. Assuming $\alpha q_{M_1} < 0$, it is $v_1 > v_2, v_3$ such that for $z < 0$, the amplitude c_1 describes the population of the first band and that of the third band for $z > 0$.

The LZM system (4.9) and (4.10), derived here for the three-level system in a biased lattice, will serve below as a prototype for other resonances, such as two-level transitions and Rabi oscillations. For example, the two-level resonances formally correspond to the system (4.9) and (4.10) with, e.g. $c_3 \equiv 0$, while for Rabi oscillations in the unbiased lattice, we can set $v_j \equiv 0$.

4.3 Rabi Oscillations

In order to study Rabi oscillations triggered by Bragg resonances in optically induced hexagonal lattices without a tilt ($|\boldsymbol{\alpha}| = 0$), we modify the derivation of the LZM model described in the previous section by using $V(\boldsymbol{r}_\perp) = \Gamma E_{\text{sc}}(I_{\text{latt}})$ in (4.1). Furthermore, we use the isotropic approximation (cf. Sect. 2.2.3) and assume a harmonic modulation of the refractive index change:

$$E_{\text{sc}} = -E_0 \frac{I_0}{1 + I_0} \cos^2(\kappa y) \cos^2\left(\kappa \frac{y + \sqrt{3}x}{2}\right) \cos^2\left(\kappa \frac{y - \sqrt{3}x}{2}\right) \tag{4.11}$$

with $I_0 = \max(I_{\text{latt}})$. The corresponding lattice wave is shown in Fig. 4.2a. $\kappa = k_y = k_x/\sqrt{3}$ defines the period of the lattice in such a way that the distance between two peaks along x-direction is $d_x = 2\pi/(\kappa\sqrt{3})$, and in y-direction $d_y = 2\pi/\kappa$. Setting $v_j \equiv 0$ in (4.10), we derive the system similar to (4.9):

$$i\frac{\mathrm{d}}{\mathrm{d}z}\begin{pmatrix} c_1 \\ c_2 \\ c_3 \end{pmatrix} = \begin{pmatrix} 0 & -\omega & -\omega \\ -\omega & 0 & -\omega \\ -\omega & -\omega & 0 \end{pmatrix}\begin{pmatrix} c_1 \\ c_2 \\ c_3 \end{pmatrix} \tag{4.12}$$

where the parameter ω denotes the *Rabi frequency* and is given by

$$\omega = \frac{3\kappa^2}{32}\Gamma E_0 \frac{I_0}{1 + I_0}. \tag{4.13}$$

The system (4.12) is fully analogous to the three-level Rabi system which describes oscillations of level populations (or quantum mechanical probability amplitudes) under the influence of an incident light field (cf. for example [20]). Following this analogy, we refer to the parameter ω in (4.13) as *Rabi frequency*. Indeed, ω is proportional to the intensity of the lattice, similar to how the semi-classical Rabi frequency is proportional to the intensity of an external light field driving the transitions. However, in our case the "levels" correspond to resonantly coupled high-symmetry points in the inverse space, also created by the lattice. As we show below, the Rabi frequency (4.13) determines the *spatial* period, along the

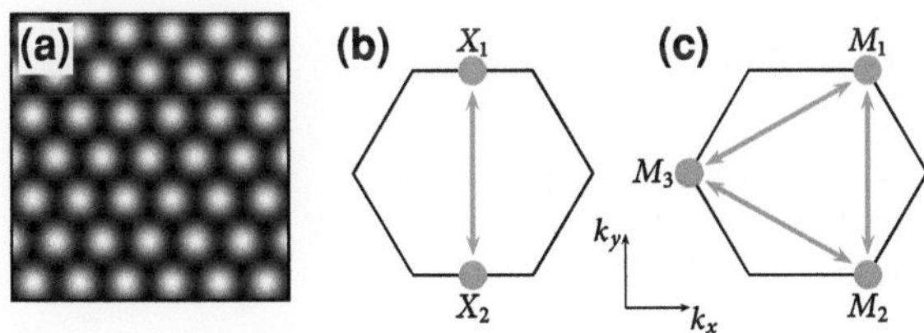

Fig. 4.2 Rabi oscillations in two-dimensional hexagonal lattices. **a** Lattice wave corresponding to (4.11); **b** two-level Rabi oscillations between two X points; **c** three-level Rabi oscillations between M points

propagation direction z, of the power oscillations between coupled high-symmetry points in Fourier space.

We distinguish two different cases of Rabi oscillations: the quasi one-dimensional oscillations shown in Fig. 4.2b which occur between the two X points and will be discussed in the next section as well as the more complex three-level Rabi oscillations between the M points (Fig. 4.2c) which we will study afterwards.

4.3.1 Two-Level Rabi Oscillations

We first consider the case of two-level Rabi oscillations between the points X_1 and X_2 as shown in Fig. 4.2b. Setting $c_3 \equiv 0$, a solution of (4.12) can be found in terms of the initial complex values $C_j = c_j(0), j = 1, 2$, namely

$$c_1 = C_1 \cos \omega z + i C_2 \sin \omega z \qquad (4.14)$$

$$c_2 = C_2 \cos \omega z + i C_1 \sin \omega z. \qquad (4.15)$$

Due to the conservation of power, i.e. $|c_1|^2 + |c_2|^2 = |C_1|^2 + |C_2|^2 = 1$, it is convenient to introduce $C_1 = \cos \varphi$ and $C_2 = \sin \varphi \exp(i\delta)$ with the initial phase difference δ and the initial power ratio $\tan^2 \varphi = |C_2|^2/|C_1|^2$. The corresponding intensities (powers) are then given by

$$|c_{1,2}|^2 = (1 \pm \cos 2\varphi \cos 2\omega z \mp \sin \delta \sin 2\varphi \sin 2\omega z)/2. \qquad (4.16)$$

In experiment, the initial conditions $C_{1,2}$ correspond to the amplitudes of two Fourier components of a composite wave launched into the lattice. Below, we investigate in detail two cases of interest. First, only one X point is excited (single-beam excitation), and second two points are excited with in-phase beams of different amplitudes or two beams having equal amplitudes but variable relative phase differences (two-beam excitation). The first approach allows direct measurements of the Rabi frequency by varying the applied voltage while the second case provides an interesting mechanism for all-optical switching.

4.3.1.1 Single-Beam Excitation

For the simplest case of single beam excitation, we have $C_1 = 1$ and $C_2 = 0$ so that $\varphi = 0$. Therefore, (4.16) transforms into the typical solution for Rabi oscillations known from two-level quantum systems:

$$|c_1|^2 = \cos^2 \omega z \qquad (4.17)$$

$$|c_2|^2 = \sin^2 \omega z. \qquad (4.18)$$

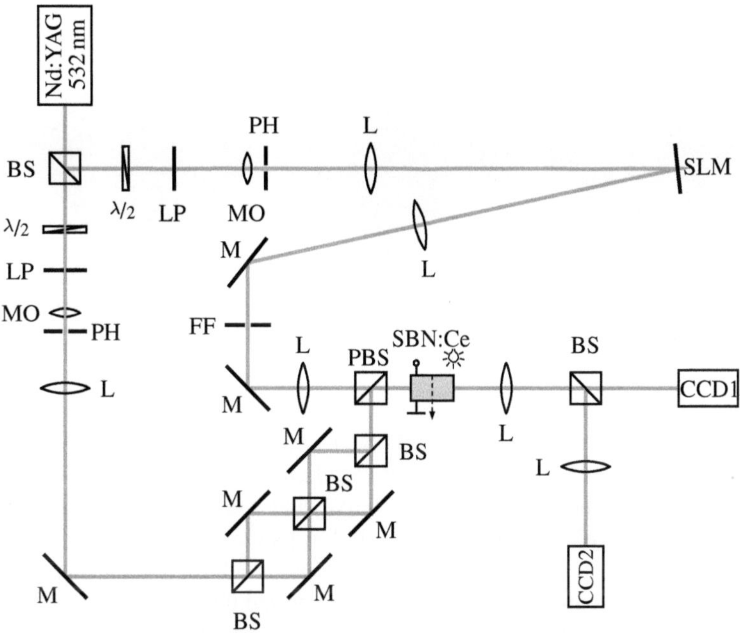

Fig. 4.3 Experimental setup for the observation of Rabi oscillations in optically induced hexagonal lattices. CCD1: real space camera; CCD2: Fourier space camera; FF: Fourier filter; L: lens; LP: linear polarizer; M: mirror; MO: microscope objective; (P)BS: (polarizing) beam splitter; PH: pinhole; SLM: spatial light modulator

For a fixed propagation length $z = L$, the output power ratio of the two Fourier peaks is given by $|c_2|^2/|c_1|^2 = \tan^2 \omega L$ and hence $\omega L = \tan^{-1}(\pm\sqrt{|c_2|^2/|c_1|^2})$. Together with (4.13), this implies that by changing the lattice intensity I_0 or the applied electric field E_0 and analyzing the output in Fourier space, we can directly measure the Rabi frequency in experiment. The setup employed for this measurement and all following measurements on Rabi oscillations in hexagonal lattices is shown schematically in Fig. 4.3.

To create the lattice beam, we use the same configuration of Mach-Zehnder type interferometers as in Fig. 3.2, whereas the probe beam is generated by illuminating a spatial light modulator and imaging the modulated beam onto the front face of a 23 mm long SBN crystal (cf. also Fig. 3.3). As before, the output of the crystal can be analyzed in real space as well as in Fourier space using two CCD cameras. For single-beam excitation, the modulator simply acts as a mirror and the lattice is probed with a Gaussian beam. In order to have a narrow Fourier spectrum around the Bragg resonance, the beam should be broad in real space and cover at least several lattice periods. Here, we choose a Gaussian beam with a full width half maximum of 320 μm and a lattice constant of $d_x \approx 22$ μm. To compensate for the photorefractive anisotropy and achieve a refractive index modulation closer to

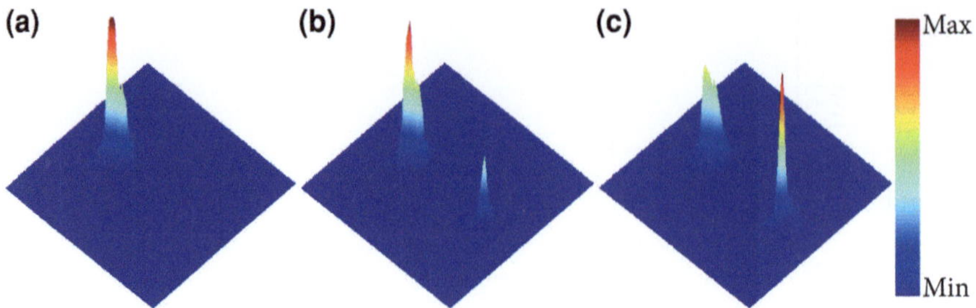

Fig. 4.4 3D visualization of the Fourier space outputs for the two-level Rabi system with single-beam excitation. (**a**)–(**c**) $E_0 = 0.2, 0.6,$ and $1\,\text{kV/cm}$, respectively

the original hexagonal symmetry (cf. Sect. 3.3), the lattice is stretched with a stretching factor $\eta = 2.4$ such that $d_y \approx 53\,\mu\text{m}$. The total power of the three interfering lattice beams is $P_{\text{latt}} = 0.3\,\text{mW}$ and the crystal is biased with an externally applied electric field of $E_0 = 0.2 - 1\,\text{kV/cm}$.

For the actual measurement, the input beam is adjusted such that its Fourier spectrum is located around the X_1 point. The applied field is then varied in steps of $\Delta E_0 = 0.2\,\text{kV/cm}$ and the output in Fourier space is analyzed by integrating the powers of the two beams at X_1 and X_2, respectively. Figure 4.4 exemplary shows the obtained output intensity distribution for three different values of the electric field. For a low field of $E_0 = 0.2\,\text{kV/cm}$, the energy transfer from the original beam at the X_1 point (strong peak in Fig. 4.4a) to the X_2 point is hardly visible, while increasing the electric field results in an increased efficiency. For $E_0 = 1\,\text{kV/cm}$, the peak at the initially unpopulated point X_2 becomes larger than the peak at X_1. In order to determine the Rabi frequency from our experiments, we use the relation

$\omega L = \tan^{-1}(\pm\sqrt{|c_2|^2/|c_1|^2})$ to calculate ωL from the integrated powers in Fourier space and plot the result versus the applied electric field as shown in Fig. 4.5. The error is estimated by conducting ten measurements for each value of the applied field followed by standard statistical analysis. In this case, the Rabi frequency can be determined by the slope of a linear fit to the experimental data which gives

$$\left[\frac{\omega L}{E_0}\right]_{\text{exp}} = 6.65 \cdot 10^{-6}\text{V}^{-1}. \tag{4.19}$$

Obviously, the same procedure of calculating ωL from the integrated powers in Fourier space can also be applied to the numerical data. The results for the parameters $I_0 = 0.125, d_x = 22\,\mu\text{m}, \eta = 2.4, x_0 = 10\,\mu\text{m}, n_0 = 2.35,$ and $r_{\text{eff}} = 280\,\text{pm/V}$ are plotted as red squares in Fig. 4.5 and lead to

$$\left[\frac{\omega L}{E_0}\right]_{\text{num}} = 6.67 \cdot 10^{-6}\text{V}^{-1}. \tag{4.20}$$

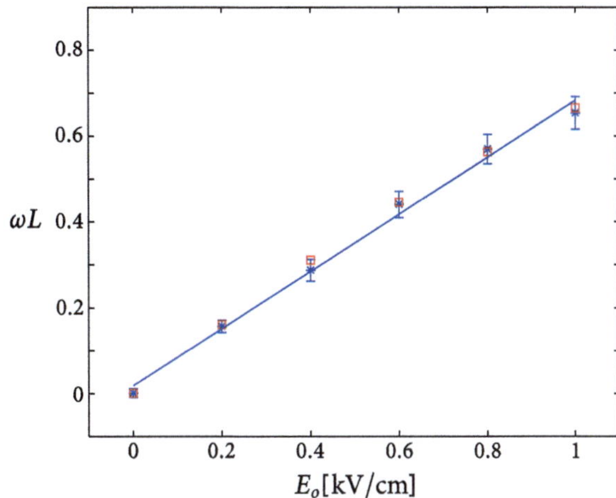

Fig. 4.5 $\omega L = \tan^{-1}(\pm\sqrt{|c_2|^2/|c_1|^2})$ as a function of E_0 for two-level Rabi oscillations with single-beam excitation. Blue markers: experimental data; red markers: numerical data; the blue line shows a linear fit to the experimental data

To compare the above results with the LZM model, we insert the numerical parameters into (4.13) and obtain

$$\left[\frac{\omega L}{E_0}\right]_{\text{LZM}} = 7.20 \cdot 10^{-6} \text{V}^{-1}. \tag{4.21}$$

This remarkable agreement between the theoretical values on the one hand and anisotropic numerics as well as experimental results on the other hand clearly proves the applicability of the LZM model to the case of optically induced photonic structures. However, it is important to note that the theory is isotropic and therefore, compensation of the photorefractive anisotropy in experiment by stretching the lattice is crucial for comparability of the results.

4.3.1.2 Two-Beam Excitation

We proceed by considering the excitation of two-level Rabi oscillations with two in-phase beams ($\delta = 0$) located at the two X points in Fourier space and analyzing the output numerically. Choosing $E_0 = 1.2\,\text{kV/cm}$ as a typical value for the applied electric field, we get $\omega L = 0.81$ from (4.20). In Fig. 4.6a, we plot the integrated output powers versus the input ratio φ as black markers (case A).

To compare the numerical results with the LZM model, we derive the corresponding solution from (4.16). Setting $\delta = 0$, we obtain

$$|c_{1,2}|^2 = (1 \pm \cos 2\varphi \cos 2\omega z)/2, \quad \sin^2\varphi \le |c_{1,2}|^2 \le \cos^2\varphi \tag{4.22}$$

where we have assumed for simplicity that $|C_1| \ge |C_2|$ so that $0 \le \varphi \le \pi/4$. The case of single beam excitation is at the limit $\varphi = 0$. With increase of φ the amplitude of the Rabi oscillations vanishes and for equal powers of two input

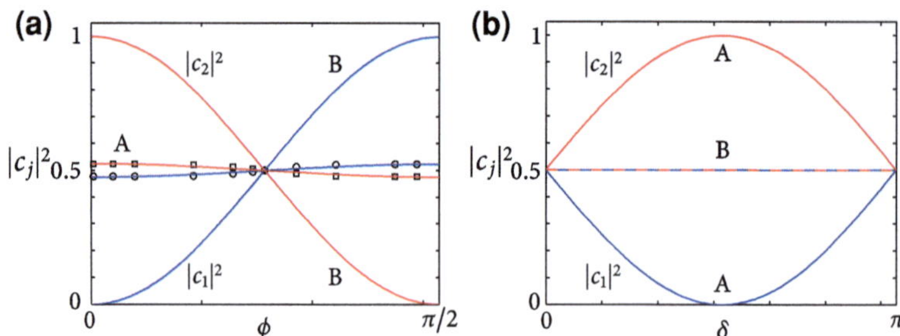

Fig. 4.6 Two-level Rabi oscillations with two-beam excitation. Lines and dots are marked A for $\omega L = 0.81$ and B for $\omega L = \pi/2$; **a** Output for $\delta = 0, \pi$, solid lines: LZM solution; markers: numerical simulation; **b** Output for $\varphi = \pi/4$ (equal input powers)

beams $\varphi = \pi/4$ the solution is stationary with $|c_{1,2}|^2 \equiv 1/2$. Again, the output powers are plotted versus the input ratio φ in Fig. 4.6a. For $\omega L = 0.81$ (case A), the results are in very good agreement with the numerical simulations, but the maximum efficiency of energy transfer between the two beams is rather small $|c_2|^2/|c_1|^2 = 0.52/0.48$.

While in the theoretical model, the efficiency can be increased by increasing ω with 100% energy transfer for $\omega L = \pi/2$ (case B in Fig. 4.6), the experimental figure of merit ωL for our system is much lower than $\pi/2$, due to the short propagation length. In fact, even for higher lattice intensities I_0 or applied fields E_0, this cannot be compensated.

In spite of this, efficient coupling between the two peaks in Fourier space can be observed in experiment by allowing the initial beams to have a nonzero phase difference $(\delta \neq 0, \pi)$. Manipulating both, powers and relative phase of the two beams sends us back to the full model [cf. (4.16)]. Instead, we separate the phase degree of freedom by considering two equal powers, $\varphi = \pi/4$, and varying the relative phase δ at the input. In this case, we obtain

$$|c_{1,2}|^2 = (1 \mp \sin \delta \sin 2\omega z)/2. \tag{4.23}$$

Remarkably, as seen in Fig. 4.6b, the efficiency is reversed with respect to the previous case in the sense that for the lower value of $\omega L = 0.8084 \simeq \pi/4$ (case A), the energy is fully transferred to c_2 (at maximal efficiency for $\delta = \pi/2$), while for larger $\omega L = \pi/2$ (case B) the solution is simply $|c_{1,2}|^2 = 0.5$ with no regard to the phase δ.

For the corresponding experiment, we use the same setup as for the single-beam excitation and employ the spatial light modulator to create two beams having their Fourier spectra located around the points X_1 and X_2 (cf. Fig. 4.7c). In real space, this configuration produces an interference pattern with a period that matches the vertical lattice constant $d_y \approx 53 \, \mu$m and varying the relative phase between the two beams results in a spatial shift of the interference fringes. This is demonstrated in Fig. 4.7a, b for a relative phase of $\delta = 0$ and $\delta = \pi$, respectively.

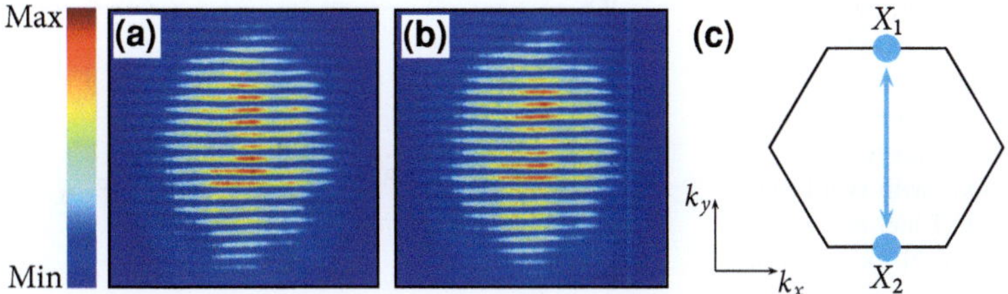

Fig. 4.7 Two-beam excitation of two-level Rabi oscillations. **a**, **b**, Real space intensity distributions for $\delta = 0$ and $\delta = \pi$, note the relative shift of the interference fringes; **c** schematic illustration of Rabi oscillations between the two X points of the first Brillouin zone

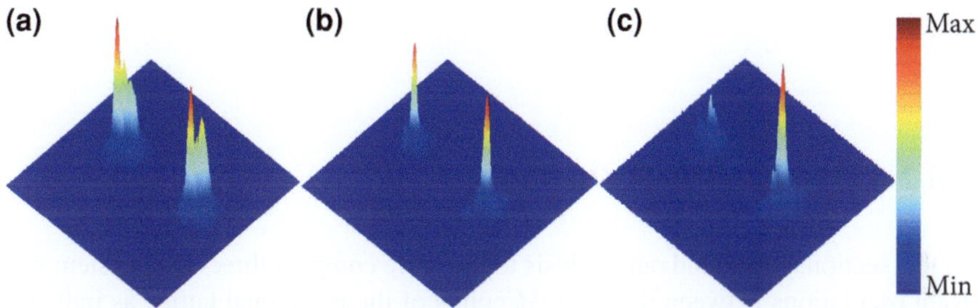

Fig. 4.8 3D visualization of the Fourier space intensity distributions for the two-level Rabi system with two-beam excitation. **a** Input; **b** output for $\delta = 0$; **c** output for $\delta = \pi/2$

As before, the total power of the three interfering lattice beams is $P_{\text{latt}} = 0.3\,\text{mW}$, but this time, we apply a fixed bias field of $E_0 = 1\,\text{kV/cm}$. The relative phase of the two beams is varied using the phase imprinting technique [cf. Appendix C] and the output is monitored in Fourier space.

Figure 4.8 shows the Fourier space intensity distributions of the input (Fig. 4.8a) as well as the two distinct cases of zero phase difference $(\delta = 0)$ with no energy transfer and a phase difference of $\delta = \pi/2$ showing maximum efficiency. As it can be seen in Fig. 4.8c this maximum efficiency in experiment is less than the 100% derived from the LZM model (cf. case A in Fig. 4.6b). This difference can be ascribed to the finite width of the experimental beams in Fourier space which reduces the effective propagation length to the overlapping region of the two beams. Nevertheless, by integrating the powers in Fourier space, we qualitatively confirm the theoretical prediction with no energy transfer for $\delta = 0, \pi$ in contrast to efficient coupling for $\delta = \pi/2$ as shown in Fig. 4.9. Similar to the single-beam excitation, the experimental error has been estimated by ten repeated measurements for each phase difference δ.

In our numerical simulations, we use the same parameters as in the case of single-beam excitation for $E_0 = 1\,\text{kV/cm}$ and observe again an excellent agreement between the numerical results and the experimental observations (Fig. 4.9).

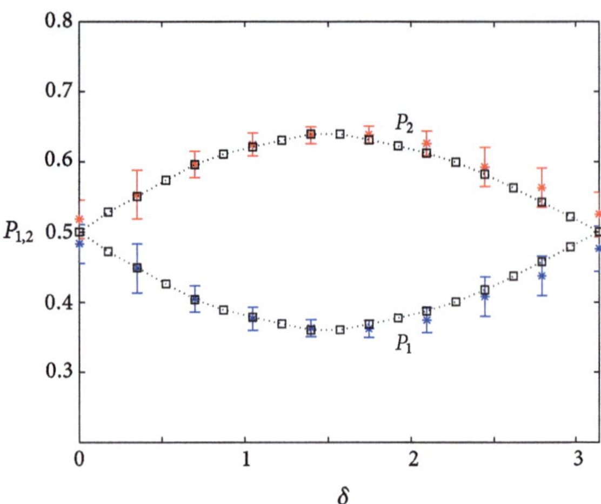

Fig. 4.9 Integrated powers in Fourier space vs. the relative phase difference δ of the two input beams for the case of two-level Rabi-oscillations with two-beam excitation. Blue, red: experiment; black: numerics; P_1 denotes the integrated power at the point X_1 and P_2 the corresponding power at X_2. Both values have been normalized to the total power $P_{\text{tot}} = P_1 + P_2$. The dotted line serves as a guide to the eye

4.3.2 Three-Level Rabi Oscillations

In this section, we extend our analysis to the more complex three-level system with Rabi oscillations between the three M points of the hexagonal lattice as indicated in Fig. 4.2c. Coupling all three M points, the general solution to (4.12) is given by

$$c_j = C_j \exp(-i\omega z) + \frac{2i}{3}(C_1 + C_2 + C_3) \sin(3\omega z/2) \exp(i\omega z/2) \tag{4.24}$$

with $j = 1, 2, 3$ and the following integrals of motion: $\sum_{j=1}^{3} |c_j|^2 = \sum_{j=1}^{3} |C_j|^2 = 1$, $|\sum_{j=1}^{3} c_j| = |\sum_{j=1}^{3} C_j|$, and $|c_m - c_n| = |C_m - C_n|$. Considering first the simple case of single-beam excitation, i.e. using initial conditions $C_1 = 1$ and $C_2 = C_3 = 0$, this solution can be reduced to

$$|c_1|^2 = (5 + 4\cos 3\omega z)/9 \tag{4.25}$$

$$|c_2|^2 = |c_3|^2 = 2(1 - \cos 3\omega z)/9. \tag{4.26}$$

Obviously, this case is similar to the two-level system in the sense that the secondary beams together form a second level with the transferred power equally divided between them. In the following, we will therefore concentrate on the more interesting case of two-beam excitation and set $C_3 = 0$ in (4.24). Similar to the two-level system above, we introduce the initial power ratio $\tan^2 \varphi = |C_2|^2/|C_1|^2$ and the phase difference δ and write $C_1 = \cos \varphi$ and $C_2 = \sin \varphi \exp(i\delta)$. Among many possible scenarios, the most interesting one is the case of two beams having equal powers ($\varphi = \pi/4$) and a variable relative phase difference δ, leading to

$$|c_{1,2}|^2 = \left[7 - 2\cos \delta + 4\cos^2(\delta/2)\cos(3\omega z) \mp 6\sin \delta \sin(3\omega z)\right]/18 \tag{4.27}$$

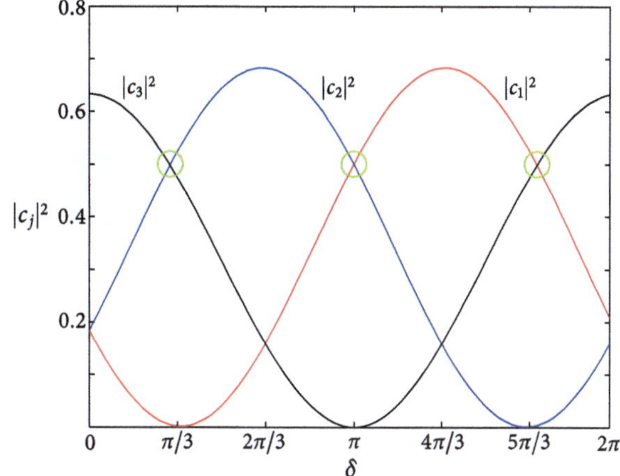

Fig. 4.10 Solution to the LZM model for three-level Rabi oscillations with two-beam excitation ($\varphi = \pi/4, \omega L = 0.67$) as a function of the relative phase difference δ; green circles mark the distinct points with only two beams at the output

$$|c_3|^2 = \frac{8}{9}\cos^2(\delta/2)\sin^2(3\omega z/2). \tag{4.28}$$

The first thing to notice here is that for any ωL the third beam c_3 disappears at $\delta \to \pi$ and the energy gets locked in two beams only. In order to connect the theoretical model (4.27) and (4.28) to the experimental situation, we choose $\omega L = 0.67$ as derived from our previous experiments on two-level Rabi oscillations with single-beam excitation (Fig. 4.5). The corresponding solution is plotted in Fig. 4.10 and shows that, apart from locking the energy in c_1 and c_2 at $\delta = \pi$, varying the relative phase allows to switch between any combination of two output beams together with a vanishing third beam (green circles in Fig. 4.10). In particular, we obtain $c_1 = 0$ for $\delta \approx \pi/3$, $c_3 = 0$ for $\delta = \pi$ and $c_2 = 0$ for $\delta \approx 5\pi/3$. Moreover, it can be seen that for $\delta = 0, 2\pi/3, 4\pi/3$, the system (4.27) and (4.28) with $\omega L = 0.67$ shows one dominating Fourier peak. As a result, varying the relative phase δ enables switching between effectively one, two as well as three beams at the output.

To observe this switching process in experiment, we generate a two-beam input having its Fourier components located at the points M_1 and M_2 as shown in Fig. 4.11b. In real space, this configuration again results in an interference pattern matching the vertical lattice constant d_y (Fig. 4.11a). The experimental parameters are chosen as before and the output is analyzed in Fourier space while varying the relative phase difference δ of the two beams at the input. Figure 4.12 shows the output intensity distributions in Fourier space for $\delta = 0, \delta = \pi/3, \delta = \pi$, as well as $\delta = 5\pi/3$ and reveals an excellent agreement with the theoretical predictions, clearly proving the switching between the expected outputs for the different relative phases. For $\delta = 0$, we obtain one dominating peak at the M_3 point (Fig. 4.12a) while the nonzero phase differences (Figs. 4.12b–d) provide all three combinations of two beams at the output together with a vanishing third beam.

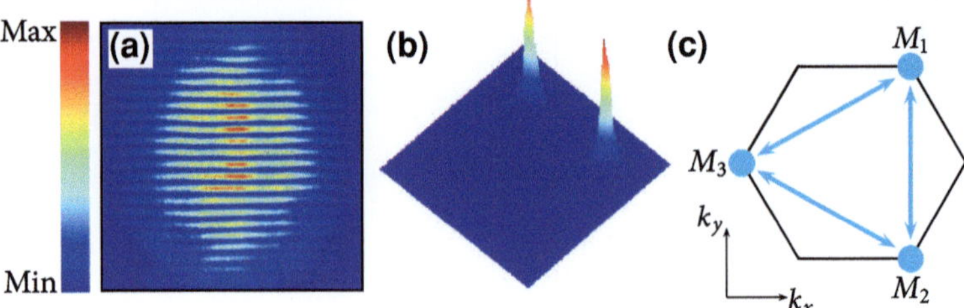

Fig. 4.11 Two-beam excitation of three-level Rabi oscillations. **a** Real space input intensity distribution; **b** input intensity distribution in Fourier space; **c** schematic illustration of Rabi oscillations between the *M* points of the first Brillouin zone

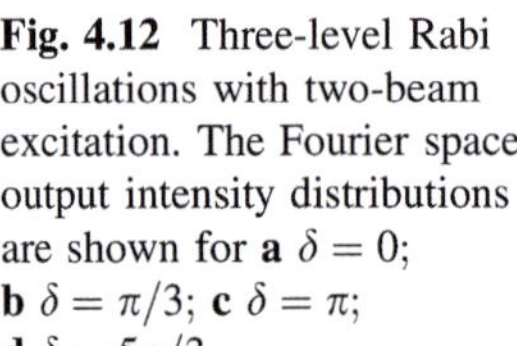

Fig. 4.12 Three-level Rabi oscillations with two-beam excitation. The Fourier space output intensity distributions are shown for **a** $\delta = 0$; **b** $\delta = \pi/3$; **c** $\delta = \pi$; **d** $\delta = 5\pi/3$

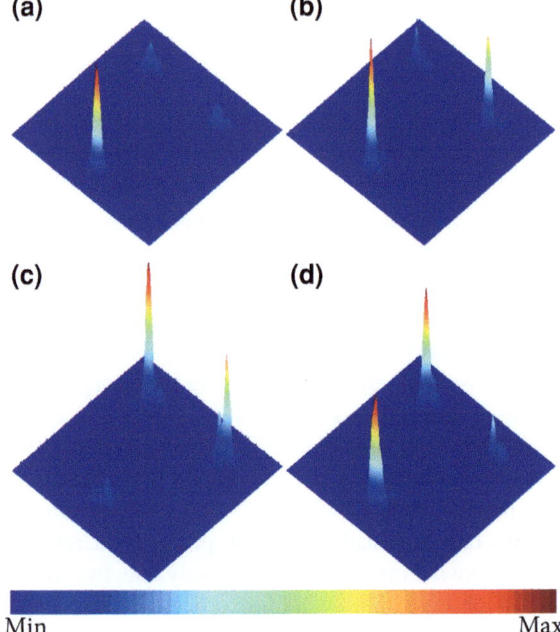

4.4 Landau-Zener Tunneling

While the Bragg resonance induced Rabi oscillations can be obtained as solutions to the modified LZM model without the linear index gradient ($|\alpha| = 0$), the process of Zener tunneling is associated with a tilted potential ($|\alpha| \neq 0$) and therefore requires consideration of the full model (4.9). A detailed theoretical analysis of this system can be found in [17].

Here, we focus on the experimental observation of Zener tunneling in hexagonal lattices accompanied by numerical simulations using the anisotropic

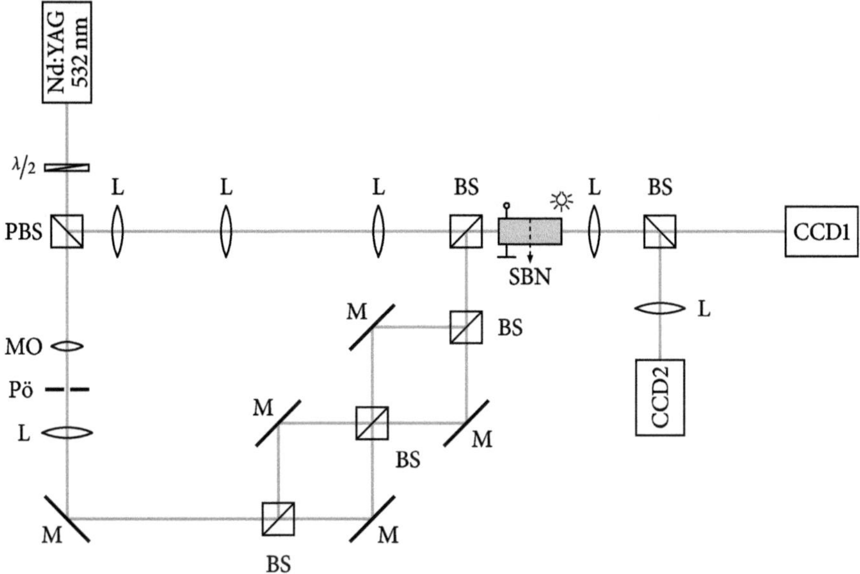

Fig. 4.13 Experimental setup for the observation of Zener tunneling in optically induced hexagonal lattices. CCD1: real space camera; CCD2: Fourier space camera; FF: Fourier filter; L: lens; M: mirror; MO: microscope objective; (P)BS: (polarizing) beam splitter; PH: pinhole

photorefractive model. To this end, we use the setup shown in Fig. 4.13. As before, the lattice wave is created by sending a laser beam through a configuration of two subsequent Mach-Zehnder type interferometers. Throughout this section, the total power of the three interfering lattice beams is chosen to be $P_{\mathrm{latt}} = 0.4\,\mathrm{mW}$ and the crystal is biased with an externally applied electric field of $E_0 = 2\,\mathrm{kV/cm}$. The lattice constants are $d_x = 24\,\mu\mathrm{m}$ and $d_y = 57\,\mu\mathrm{m}$, thus giving a stretching factor of $\eta \approx 2.4$. The probe beam is formed by a Gaussian beam covering several lattice sites at the front face of the same 23 mm long photorefractive SBN crystal that has already been used for the observation of Rabi oscillations in the previous section.

In order to induce a refractive index gradient along the transverse x-direction, the crystal is illuminated from the top with a modulated incoherent white light intensity distribution. The modulation is achieved by placing a mask directly on top of the crystal and utilizing the shade of this mask while it is illuminated. To get an impression of the resulting intensity gradient inside the crystal, the transverse intensity distribution in a plane 2 mm behind the mask has been measured in a separate experiment. The result is depicted in Fig. 4.14 and shows an approximately linear dependence in the central region of the gradient. The performance of this gradient has been tested by launching a probe beam in the resulting structure without a superimposed lattice. As expected, we have found that the beam experiences a transverse shift which depends on the applied electric field. For $E_0 = 2\,\mathrm{kV/cm}$, as used for the experiments in this section, we found a shift of

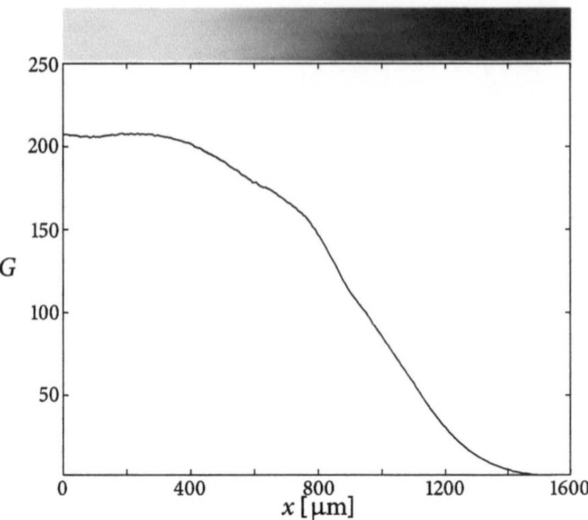

Fig. 4.14 Experimentally measured transverse profile of the modulated background illumination. (*Top*) CCD image; (*bottom*) plot of gray value G vs. spatial coordinate x

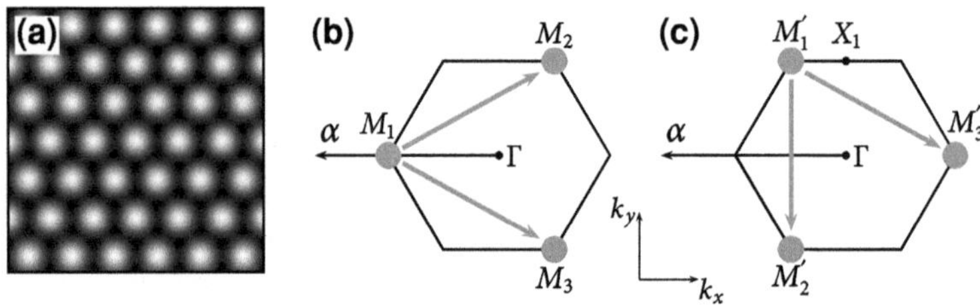

Fig. 4.15 Zener tunneling at the symmetric threefold resonance (M points) of the hexagonal lattice. **a** Lattice wave; **b** symmetric tunneling; **c** asymmetric tunneling

approximately $100\,\mu$m. In contrast to the theoretical analysis, in which the gradient can be oriented along any direction [17], the above technique only allows to produce a refractive index gradient along the x-direction. In addition, the lattice wave has to be oriented in the horizontal configuration to minimize the influence of anisotropy on the induced refractive index structure [cf. Sect. 3.3]. Therefore, the index gradient in our experiments is always directed along the horizontal direction of the first Brillouin zone (Fig. 4.15).

Nevertheless, the experimental system still enables the investigation of two interesting cases related to the threefold resonance at the M points. The first case is shown in Fig. 4.15b. In presence of the gradient, a probe beam with initial condition $k_y = 0$ moves across the Brillouin zone before reaching the Bragg resonance (point M_1 in Fig. 4.15b). Eventually, two additional beams appear due to the resonant coupling between the Fourier amplitudes (cf. Sect. 4.2). As the transferred energy is equally distributed between those additional beams, we will refer to this case as *symmetric tunneling* in the following.

In the second situation, the beam moves along the *XM*-line of the first Brillouin zone, thus reaching the Bragg resonance at $k_y \neq 0$ (point M_1' in Fig. 4.15c). Due to the initial symmetry breaking, the energy is generally not distributed equally between the appearing beams [points M_2' and M_3' in Fig. 4.15c] and we will denote this case as *asymmetric tunneling*. Since it is not possible to directly observe the evolution of the probe beam inside the crystal in experiment, we vary the incident angle and image the real space as well as the Fourier space output at the back face of the crystal. For a fixed crystal length and angles below the Bragg resonance, such an excitation at different transverse wave vector components is equivalent to different starting points in the Brillouin zone and thus allows to infer details of the tunneling dynamics at different stages of the beam evolution [12].

4.4.1 Symmetric Tunneling

The results for the case of symmetric tunneling are summarized in Fig. 4.16. Due to the index gradient, the beam moves across the lattice and eventually part of the initial beam energy tunnels into two different channels as can clearly be seen in the real space pictures of Fig. 4.16. To get more detailed information about the tunneling process, we monitor the spatial Fourier spectrum of the output field. In Fourier space, the refractive index gradient results in a movement of the beam through the Brillouin zone along the ΓM-line. When it reaches the M point, two additional beams appear. However, in full agreement with the theoretical studies [16, 17], we find that the final efficiency of the tunneling process strongly depends on the initial conditions, i.e. on the initial inclination angle of the probe beam. This is demonstrated in the Fourier space pictures of Fig. 4.16 where the intensity of the tunneled beams cleary varies with the input angle. In fact, we monitor an oscillatory behavior ranging from weak tunneled beams for a very small input angle $\alpha = 0.102°$ (Fig. 4.16b) over almost equal intensities of all three beams in Fig. 4.16d ($\alpha = 0.285°$) to very strong tunneled beams, i.e. high tunneling efficiency for an input angle $\alpha = 0.340°$ as shown in Fig. 4.16f. Increasing the angle further to $\alpha = 0.520°$ then results in weaker tunneled beams again (Fig. 4.16h).

In our numerical simulations, we approximate the profile of the background illumination by

$$I_{\mathrm{bg}} = \frac{B}{2}\left[1 + \tanh\left(\frac{-x}{20}\right)\right] \tag{4.29}$$

and solve (3.1) using $I_{\mathrm{tot}} = I_{\mathrm{latt}} + I_{\mathrm{bg}}$. In analogy to the experimental situation, we model the beam propagation for different input angles and analyze the output in real space as well as in Fourier space. As shown in Fig. 4.16, the numerical results for both cases are in good qualitative agreement with the experiments.

As discussed in Sect. 4.2, tunneling of the initial beam at the corresponding M point [M_1 in Fig. 4.15b] happens between the first and the third Bloch band.

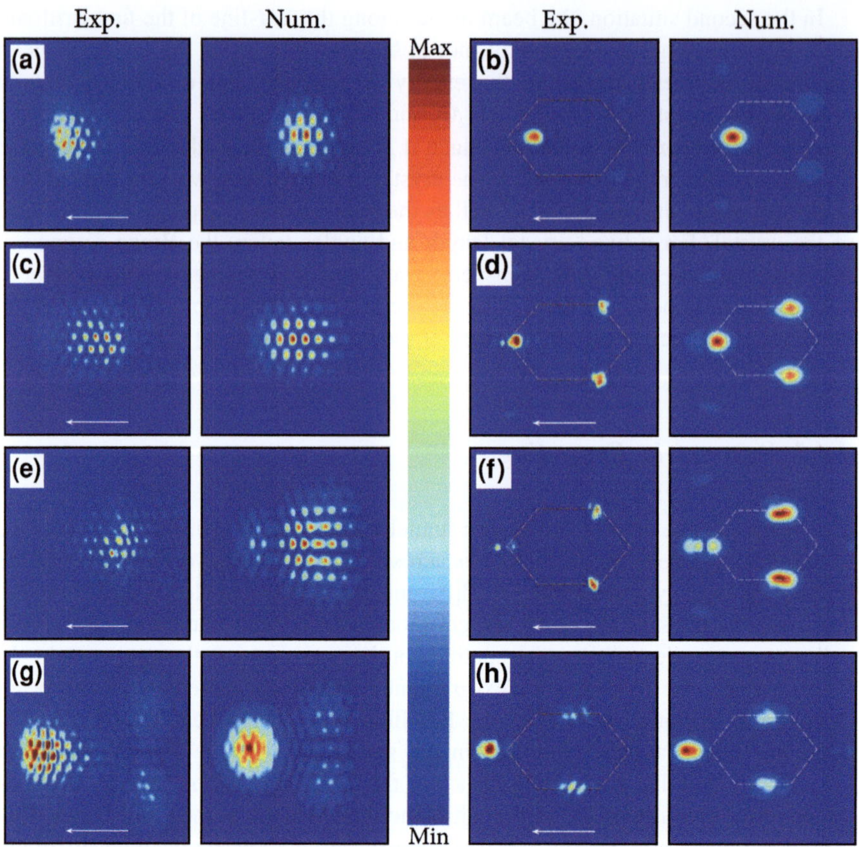

Fig. 4.16 Symmetric tunneling: real space (*left*) and Fourier space (*right*) output intensity distribution for different initial inclination angles θ of the probe beam. White arrows indicate the direction of the refractive index gradient. The first Brillouin zone in the Fourier space pictures is marked by dashed hexagons. **a, b** $\theta = 0.102°$; **c, d** $\theta = 0.285°$; **e, f** $\theta = 0.340°$; **g, h** $\theta = 0.520°$

Accordingly, the real space outputs in Fig. 4.16g show regions of high intensity located between the lattice sites which can be seen as a clear indication of a Bloch wave belonging to a higher band [21]. To get a more qualitative measure of how the efficiency depends on the inclination angle, we proceed in a similar way as in the case of Rabi oscillations and integrate the powers of the three beams at the M points of the first Brillouin zone. A comparison of the numerical values with those obtained from the experimental data is shown in Fig. 4.17. Again, we notice a remarkable agreement between numerics and experiment, clearly confirming the oscillatory character of the tunneling efficiency.

Another interesting fact to note from Figs. 4.16 and 4.17 is that while crossing the M point, the resonant tunneling results in a three-beam interference which reproduces the wave used to optically induce the lattice itself. This is demonstrated

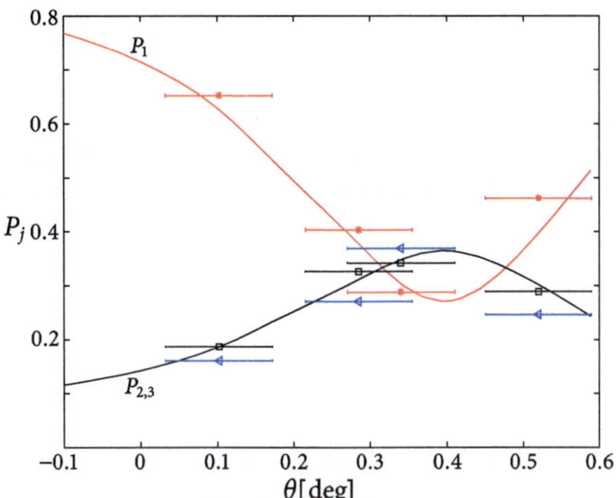

Fig. 4.17 Integrated powers of the input beam P_1 (red) and the two tunneled beams $P_{2,3}$ (blue, black) for the case of symmetric resonance. Solid lines show data obtained from numerical simulations. The angular error of the experimental data (markers) is due to the finite width of the beams in Fourier space. P_j denotes the integrated power at the point $M_j(j = 1, 2, 3)$, normalized to the total power $P_{\text{tot}} = P_1 + P_2 + P_3$

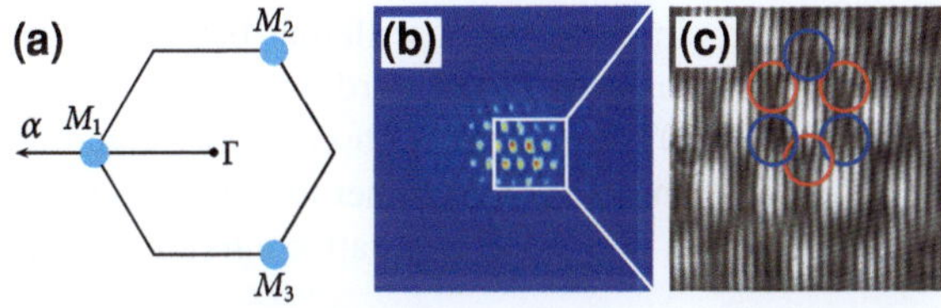

Fig. 4.18 Bloch wave generation by symmetric tunneling. **a** Schematic illustration of the first Brillouin zone with the three M points marked by blue dots and an index gradient along the ΓM-line; **b** intensity distribution for $\theta = 0.285°$ [cf. Fig. 4.16c]; **c** interferogram with a broad plane wave, circles indicate positions of phase dislocations

in Fig. 4.18 where we have monitored the phase structure of the real space output from Fig. 4.16c by interference with a plane wave and observed a regular phase structure with an imbedded honeycomb lattice of optical vortices which we have already seen in Sect. 3.3. In experiment, the vortices appear as characteristic forks in the interferogram and have been marked by red and blue circles in Fig. 4.18c. As such a honeycomb pattern of phase singularities is a signature of the corresponding Bloch wave at the M point of the hexagonal lattice [22], we conclude that Zener tunneling can be employed for the excitation of specific Bloch waves and thus for the characterization of periodic photonic structures.

4.4.2 Asymmetric Tunneling

We continue by considering the asymmetric tunneling as indicated in Fig. 4.15c. The probe beam is initially placed at the X point of the Brillouin zone [point X_1 in

Fig. 4.15c] and follows the gradient along the XM-line. As shown in [17], the analysis of the LZM model for this case reveals that before and after reaching the threefold resonance at the M point, the situation is similar to the two-level Rabi oscillations discussed in Sect. 4.3.1, i.e. the Fourier amplitudes c_1 and c_2 form an oscillating state governed by the system (4.14) and (4.15). In this case, the coupling to the third Fourier amplitude c_3 is non-resonant and can be neglected. Eventually, the oscillating state reaches the resonance at the M point where part of the energy is transferred to c_3 such that the amplitude of oscillation gets reduced. The relation of this behavior to the band structure shown in Fig. 4.1 is the following: before the resonance, the Rabi oscillations are between the first and second band (along the XM-line) and continue between the second and third band after the resonance. Therefore, the case of asymmetric tunneling has also been denoted as *tunneling of Rabi oscillations* [17]. For the experimental observation of this process, we use the same setup as before (Fig. 4.13) and monitor again the real space and Fourier space output while varying the inclination angle of the probe beam. The results for two different angles are shown in Fig. 4.19. For $\theta = 0.017°$, we observe a state of the Rabi oscillation in which the initial beam is stronger than the second one $(P_1 > P_2)$ and since the two oscillating beams do not cross the resonance, there is hardly any coupling to the third beam (Fig. 4.19a, b). Increasing the input angle to $\theta = 0.290°$, we clearly observe this coupling as well as a different state of the oscillation between the other two beams $(P_1 < P_2)$. As before, our experimental results are in very good qualitative agreement with the numerical simulations.

Similar to the symmetric case, a more quantitative measure of the tunneling efficiency is obtained by integrating the powers of the respective Fourier peaks from the experimental data and the numerical simulations. The obtained results

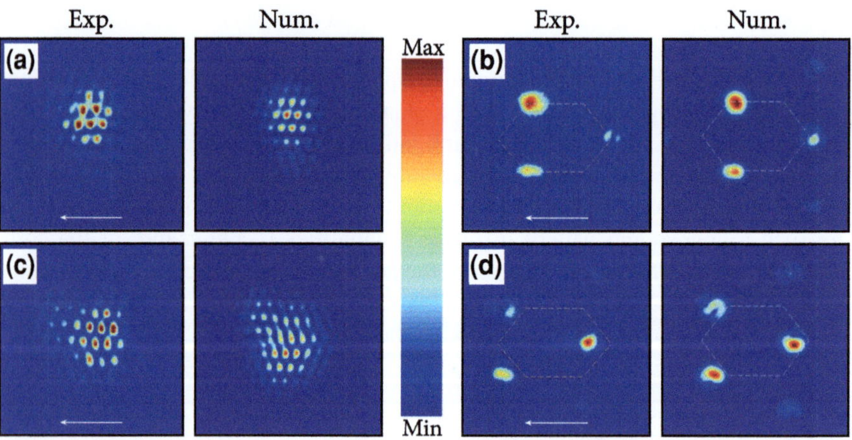

Fig. 4.19 Asymmetric tunneling: Real space (*left*) and Fourier space (*right*) output intensity distribution for different initial inclination angles θ of the probe beam. White arrows indicate the direction of the refractive index gradient. The first Brillouin zone in the Fourier space pictures is marked by dashed hexagons. (**a**), (**b**) $\theta = 0.017°$; (**c**), (**d**) $\theta = 0.290°$

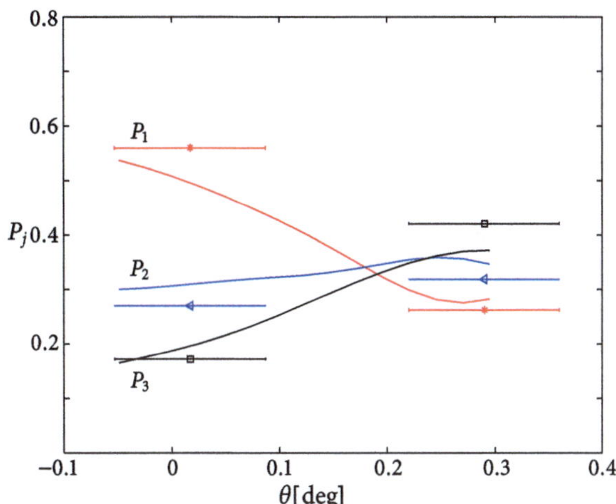

Fig. 4.20 Integrated powers of the input beam P_1 (red) and the two tunneled beams $P_{2,3}$ (blue, black) for the case of asymmetric resonance. Solid lines show data obtained from numerical simulations. The angular error of the experimental data is due to the finite width of the beams in Fourier space. P_j denotes the integrated power at the point $M_j (j = 1, 2, 3)$, normalized to the total power $P_{\text{tot}} = P_1 + P_2 + P_3$

shown in Fig. 4.20 confirm the oscillation between P_1 and P_2 as well as the energy transfer to P_3 which results in a reduced amplitude of oscillation.

4.5 Nonlinear Tunneling

So far, any influence of nonlinearity on the previously discussed resonant processes has been neglected. However, such effects have been shown to play an important role in the dynamics of Bose-Einstein condensates in optical lattices, because in most instances the nonlinearity due to atom-atom interaction cannot be neglected (cf. Review in [23]). Probably the simplest arrangement in the studies of nonlinear tunneling of Bose-Einstein condensates is the double-well potential in which the nonlinearity leads to periodic oscillations of atoms between the wells and macroscopic quantum self-trapping [24, 25].

The reason for such dramatic effects is that, since the superposition principle is no longer valid in the presence of nonlinearity, the linear modes of the potential are effectively coupled. Similarly, the Bloch waves fail to represent uncoupled modes of the lattice and start to interact with each other, i.e. the nonlinearity effectively couples different bands and leads to significant modifications of the wave dynamics. In particular, for the case of one-dimensional accelerated (biased) lattices, the nonlinearity introduces an asymmetry in the sense that tunneling from the ground state to the excited state gets enhanced, whereas tunneling in the opposite direction becomes suppressed [26].

Nonlinear tunneling was studied theoretically in two-dimensional square lattices and nonlinear LZM systems, similar to (4.9) and (4.10), have been derived [27–29]. The main modification to the linear LZM system is the appearance of additive nonlinear terms in the diagonal elements of the matrix H in (4.10). This is similar to cross-phase modulation which couples different high-symmetry points even without bias. However, as we demonstrate below, such systems cannot adequately describe the dynamics observed in experiments, because they only involve the levels which are resonantly coupled in the linear limit. In contrast, depending on the excitation, the nonlinearity leads to tunneling of optical power to additional high-symmetry points.

In experiments with photonic lattices in photorefractive media, the transition to the nonlinear regime is achieved by simply increasing the power of the input beam. This way, we have performed two series of experiments in order to investigate nonlinear Rabi oscillations as well as nonlinear Landau-Zener tunneling. In both cases, we use the same setup and lattice parameters as in the corresponding linear experiments described in Sects. 4.3 and 4.4. The results are presented in Figs. 4.21 and 4.22, respectively.

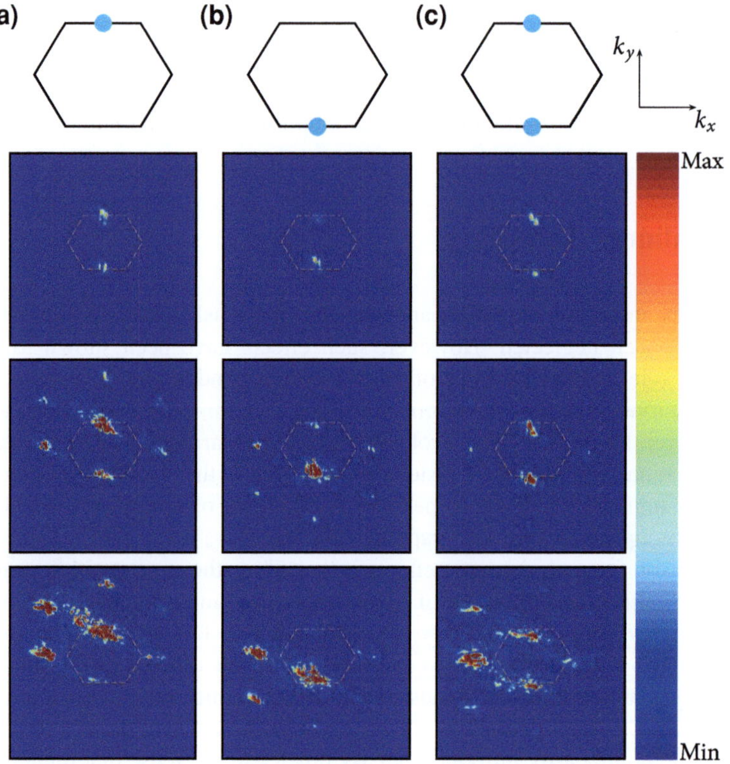

Fig. 4.21 Nonlinear interband tunneling in the unbiased lattice. (**a**), (**b**) Single-beam excitation; (**c**) two-beam excitation. The input is shown schematically in the top panels. In the color images of the output, the power is increased from top to bottom for all three cases. For low input power ($\approx 0.2\,\mu$W), single-beam as well as two-beam excitations at the X points repeat the results of two-level Rabi oscillations, transferring the input to the two output Fourier components (*second row*). With an increase of power to $\approx 0.6\,\mu$W, the energy is transferred to the high-symmetry points which are closest to the excitation point (*third row*). A further increase of power to $\approx 1\,\mu$W leads to modulational instability (*bottom row*)

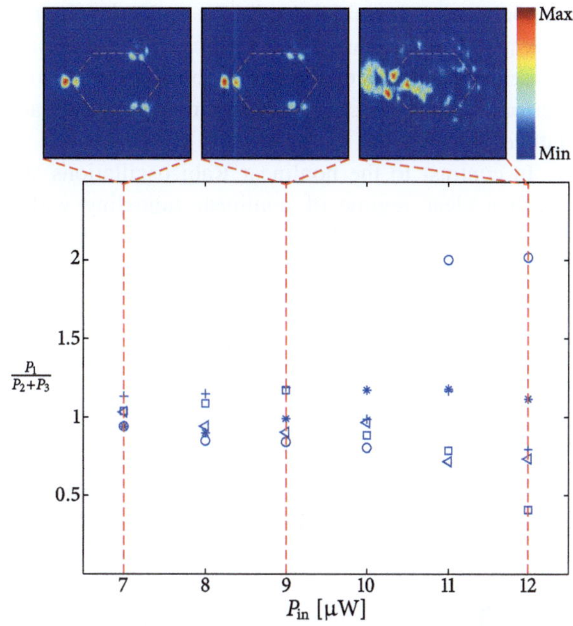

Fig. 4.22 Nonlinear interband tunneling in the biased hexagonal lattice. With an increase of input power, there is a sharp transition between linear Landau-Zener tunneling (up to $\approx 10\,\mu\text{W}$) and modulational instability which is clearly seen in the Fourier image at a power of $\approx 12\,\mu\text{W}$ (*top right*)

To realize the most accessible system, we first consider the simplest two-level nonlinear Rabi oscillations and summarize our results in Fig. 4.21. As shown schematically in the top row of Fig. 4.21, we compare three different initial conditions for single-beam excitation (Fig. 4.21a, b) as well as two-beam excitation (Fig. 4.21c). At low power, the Fourier space output shows two excited X points in all three cases. This is in full agreement with the linear Rabi oscillations described in Sect. 4.3. With an increase of power up to $\approx 600\,\text{nW}$, the output changes dramatically and we observe the appearance of five new excited X points in the second Brillouin zone as clearly visible in Fig. 4.21. A comparison of Figs. 4.21a, b shows that, with no regard to the center of the first Brillouin zone, the power is transferred to the X points closest to the initially excited point. As a result, the two additional X points on the left and on the right from the first Brillouin zone are present in both cases, while three more X points are excited above (Fig. 4.21a) and below (Fig. 4.21b) the first Brillouin zone. This asymmetry is eliminated in the case of two-beam excitation shown in Fig. 4.21c. Here, the power is most efficiently transferred to the left and right X points while it is evenly distributed between the three top and three bottom X points which are hardly visible. A further increase of power to $\approx 1\,\mu\text{W}$ clearly shows the development of modulational instability with significantly broadened and disordered Fourier peaks (bottom row of frames in Fig. 4.21). Interestingly, we observe an asymmetry in the development of modulational instability, persistent for all three different excitations, namely the left Fourier peaks getting much stronger than the right ones.

The appearance of modulational instability and its intrinsic relation to interband tunneling was discussed in [30] for nonseparable square lattices. The crucial importance of the initial conditions to the development of modulational instability was noted in the context of asymmetry of nonlinear Landau-Zener tunneling, in agreement with our observations above. In order to fully appreciate this effect, we perform additional experiments on nonlinear Landau-Zener tunneling in the biased lattice.

In contrast to the nonlinear Rabi oscillations shown in Fig. 4.21, we did not reach a clear regime of nonlinear tunneling with excitation of high-symmetry points in the second Brillouin zone here, although the efficiency of tunneling is somewhat enhanced for intermediate input power (cf. relative powers of coupled M points for $7\,\mu$W and $9\,\mu$W in Fig. 4.22). Instead, with further increase of power, we observe a sharp transition to the modulational instability regime as shown in the Fourier image at $7\,\mu$W (top right panel). The left-to-right asymmetry is clearly present, similar to Fig. 4.21. However, here it can be assigned to the action of the transverse refractive index gradient.

To conclude, we have observed interband coupling induced by nonlinearity and identified two different regimes: the resonant nonlinear tunneling in unbiased lattices, or the so-called Bloch band tunneling [30], and the development of modulational instability in both, pure periodic and biased lattices. In addition, we observe the left-to-right asymmetry of tunneling, persistent not only in biased lattice, but also in the regime of nonlinear Rabi oscillations. Unfortunately, the data available so far are not conclusive about the reason for such an asymmetry.

References

1. Lederer, F., Stegeman, G., Christodoulides, D., Assanto, G., Segev, M., Silberberg, Y.: Discrete solitons in optics. Phys. Rep. **463**, 1 (2008)
2. Bloch, F.: Über die quantenmechanik der elektronen in kristallgittern. Z. Phys. **52**, 555 (1928)
3. Zener, C.: Non-adiabatic crossing of energy levels. Proc. R. Soc. Lond. A **137**, 696 (1932)
4. Zener, C.: A theory of electrical breakdown of solid dielectrics. Proc. R. Soc. Lond. A **145**, 523 (1934)
5. Esaki, L., Long journey into tunneling. Rev. Mod. Phys. **46**, 237 (1974)
6. Sibille, A., Palmier, J.F., Laruelle, F.: Zener interminib and resonant breakdown in superlattices. Phys. Rev. Lett. **80**, 4506 (1998)
7. Pertsch, T., Dannberg, P., Elflein, W., Bräuer, A., Lederer, F.: Optical Bloch oscillations in temperature tuned waveguide arrays. Phys. Rev. Lett. **83**, 4752 (1999)
8. Morandotti, R., Peschel, U., Aitchison, J.S., Eisenberg, H.S., Silberberg, Y.: Experimental observation of linear and nonlinear optical Bloch oscillations. Phys. Rev. Lett. **83**, 4756 (1999)
9. Trompeter, H., Pertsch, T., Lederer, F., Michaelis, D., Streppel, U., Bräuer, A., Peschel, U.: Visual observation of Zener tunneling. Phys. Rev. Lett. **96**, 023901 (2006)
10. Sapienza, R., Costantino, P., Wiersma, D., Ghulinyan, M., Oton, C.J. , Pavesi, L.: Optical analogue of electronic Bloch oscillations. Phys. Rev. Lett. **91**, 263902 (2003)
11. Agarwal, V., del Río, J.A., Malpuech, G., Zamfirescu, M., Kavokin, A., Coquillat, D., Scalbert, D., Vladimirova, M., Gil, B.: Photon Bloch oscillations in porous silicon optical superlattices. Phys. Rev. Lett. **92**, 097401 (2004)

12. Trompeter, H., Krolikowski, W., Neshev, D.N., Desyatnikov, A.S., Sukhorukov, A.A., Kivshar, Y.S., Pertsch, T., Peschel, U., Lederer, F.: Bloch oscillations and Zener tunneling in two-dimensional photonic lattices. Phys. Rev. Lett. **96**, 053903 (2006)
13. Landau, L.D.: On the theory of transfer of energy at collisions II. Phys. Z. Sowjetunion **2**, 46 (1932)
14. Majorana, E.: Orientated atoms in a variable magnetic field. Nuovo Cimento **9**, 43 (1932)
15. Shchesnovich, V.S., Cavalcanti, S.B., Hickmann, J.M., Kivshar, Y.S.: Zener tunneling in two-dimensional photonic lattices. Phys. Rev. E **74**, 056602 (2006)
16. Desyatnikov, A.S., Kivshar, Y.S., Shchesnovich, V.S., Cavalcanti, S.B., Hickmann, J.M.: Resonant Zener tunneling in two-dimensional periodic photonic lattices. Opt. Lett. **32**, 325 (2007)
17. Shchesnovich, V.S., Desyatnikov, A.S., Kivshar, Y.S.: Interb and resonant transitions in two-dimensional hexagonal lattices: Rabi oscillations, Zener tunneling, and tunneling of phase dislocations. Opt. Express **16**, 14076 (2008)
18. Shchesnovich, V.S., Chávez-Cerda, S.: Bragg-resonance-induced Rabi oscillations in photonic lattices. Opt. Lett. **32**, 1920 (2007)
19. Ashcroft, N.W., Mermin, N.D.: Solid State Physics. Holt, Rinehart and Winston, New York (1976)
20. Paschotta, R.: Encyclopedia of Laser Physics and Technology. Wiley-VCH, Berlin (2008)
21. Träger, D., Fischer, R., Neshev, D.N., Sukhorukov, A.A., Denz, C., Krolikowski, W., Kivshar, Y.S.: Nonlinear Bloch modes in two-dimensional photonic lattices. Opt. Express **14**, 1913 (2006)
22. Alexander, T.J., Desyatnikov, A.S., Kivshar, Y.S.: Multivortex solitons in triangular photonic lattices. Opt. Lett. **32**, 1293 (2007)
23. Morsch, O., Oberthaler, M.: Dynamics of Bose-Einstein condensates in optical lattices. Rev. Mod. Phys. **78**, 179 (2006)
24. Smerzi, A., Fantoni, S., Giovanazzi, S., Shenoy, S.R.: Quantum coherent atomic tunneling between two trapped Bose-Einstein condensates. Phys. Rev. Lett. **79**, 4950 (1997)
25. Albiez, M., Gati, R., Fölling, J., Hunsmann, S., Cristiani, M., Oberthaler, M.K.: Direct observation of tunneling and nonlinear self-trapping in a single bosonic Josephson junction. Phys. Rev. Lett. **95**, 010402 (2005)
26. Jona-Lasinio, M., Morsch, O., Cristiani, M., Malossi, N., Müller, J.H., Courtade, E., Anderlini, M., Arimondo, E.: Asymmetric L and au-Zener tunneling in a periodic potential. Phys. Rev. Lett. **91**, 230406 (2003)
27. Shchesnovich, V.S., Konotop, V.V.: Nonlinear tunneling of Bose-Einstein condensates in an optical lattice: signatures of quantum collapse and revival. Phys. Rev. A **75**, 063628 (2007)
28. Brazhnyi, V.A., Konotop, V.V., Kuzmiak, V., Shchesnovich, V.S.: Nonlinear tunneling in two-dimensional lattices. Phys. Rev. A **76**, 023608 (2007)
29. Wang, G.F., Ye, D.F., Fu, L.B., Chen, X.Z., Liu, J.: Landau-Zener tunneling in a nonlinear three-level system. Phys. Rev. A **74**, 033414 (2006)
30. Brazhnyi, V.A., Konotop, V.V., Kuzmiak, V.: Nature of the intrinsic relation between Bloch-b and tunneling and modulational instability. Phys. Rev. Lett. **96**, 150402 (2006)

Chapter 5
Nonlinear Light Localization

5.1 Solitons in Periodic Photonic Structures

We have already seen in Sect. 2.3.3 that light propagation in nonlinear photonic structures enables the formation of discrete and gap solitons having propagation constants located in the gaps of the linear spectrum. After their first prediction as discrete localized states in nonlinear waveguide arrays [1], the investigation of optical solitons in periodic potentials has become a very active research area in recent years [2]. In the following, we will briefly recall some important works on discrete and gap solitons in optics and subsequently focus on the existence and stability of different types of solitons with complex phase structures containing one or more phase singularities.

5.1.1 Discrete and Gap Solitons

Optical discrete solitons with propagation constants in the semi-infinite gap were first observed experimentally in one-dimensional AlGaAs waveguide arrays in 1998 (cf. Fig. 2.4) and later reproduced in one-dimensional optically induced photonic lattices [3, 4]. For the case of higher order bands, the existence of gap solitons located in the finite size Bragg reflection gaps was predicted in 1993 [5, 6].

While discrete solitons can easily be excited by a narrow input beam launched into a single lattice site, the excitation of gap solitons is less trivial. Nevertheless, they have been observed experimentally using periodically modulated input beams in waveguide arrays [7] as well as in optically induced photonic lattices [8].

The existence of gap solitons in one-dimensional systems with defocusing nonlinearity has also been predicted in 1993 [9]. Originating from the bottom of the first band at $k_x = \pi/d$, their propagation constant lies in the first gap and the Bragg condition gives them a charateristic staggered phase structure. They have

B. Terhalle, *Controlling Light in Optically Induced Photonic Lattices*, Springer Theses, DOI: 10.1007/978-3-642-16647-1_5, © Springer-Verlag Berlin Heidelberg 2011

first been observed using an inclined probe beam in optically induced lattices [4] and are particularly interesting, because they have no equivalent in homogeneous media in which diffraction would always be increased by the defocusing nonlinearity.

The localization of optical beams becomes even more interesting in two-dimensional systems. For example, routing and switching applications in discrete soliton networks have been suggested [10, 11] and it has been demonstrated that two-dimensional photonic lattices may lead to highly anisotropic mobility properties of gap solitons.

Experimentally, two-dimensional discrete and gap solitons have both been observed in optically induced photonic lattices [12–14]. As a simple example, Fig. 5.1 illustrates the formation of a fundamental discrete soliton in a two-dimensional optically induced photonic lattice of square symmetry. In this experiment, a Gaussian probe beam is focused into a single lattice site at the front face of a biased photorefractive SBN crystal and the output is monitored for three different input powers (Fig. 5.1a–c). At low input power, the beam experiences typical discrete diffraction as shown in Fig. 5.1a. When the power is increased, the nonlinearity leads to the formation of a discrete soliton as demonstrated in Fig. 5.1b, c. The experimental results are in excellent agreement with the anisotropic numerical simulations shown in Fig. 5.1d–f.

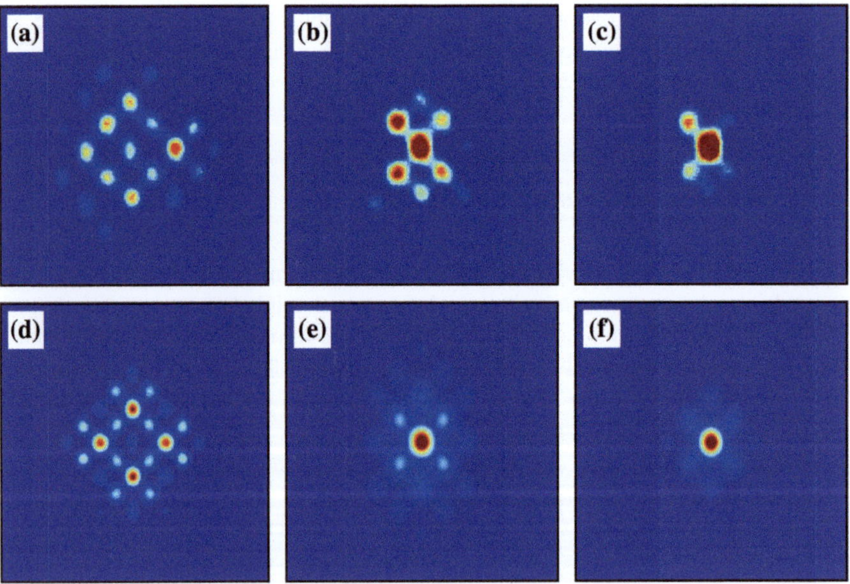

Fig. 5.1 Experimental results (*top*) and numerical simulations (*bottom*) for the formation of a fundamental discrete soliton in an optically induced diamond lattice. Results are shown for different input powers. **a, d** Discrete diffraction at 32 nW; **b, e** self-focusing at 110 nW; **c, f** discrete soliton at 250 nW

Note that the lattice is oriented in the diamond configuration, i.e. it is rotated by 45° with respect to the c-axis of the crystal. This is important since it has been demonstrated that the orientation anisotropy (cf. Sect. 3.2.3) not only determines the induced lattice structure but also strongly influences the formation of discrete solitons in optically induced photonic lattices [13]. For this reason, the majority of nonlinear localization experiments in optically induced photonic lattices has been restricted to the diamond pattern and the effects of anisotropy have been neglected. Only recently, we have considered different two-dimensional lattice geometries showing distinct anisotropic features of the induced refractive index as well as discrete and gap soliton formation [13, 15].

To give an example of such complex solitons in highly anisotropic two-dimensional photonic structures, Fig. 5.2 shows the observation of a dipole-mode gap soliton in a triangular lattice which can be considered as a higher order lattice consisting of dipole structures oriented in a diamond pattern with angles of 60°. As a result, the induced refractive index pattern for this lattice type gives rise to the formation of dipole-mode gap solitons [15].

In the corresponding experiment, a dipole-like input beam is focused to the front face of the crystal and the output is again monitored for three different input powers (Fig. 5.2a–c). At low input power, the diffraction pattern consists of a central dipole surrounded by four side lobes each forming a dipole itself (Fig. 5.2b). With increased power, the side lobes vanish and a stable dipole-mode

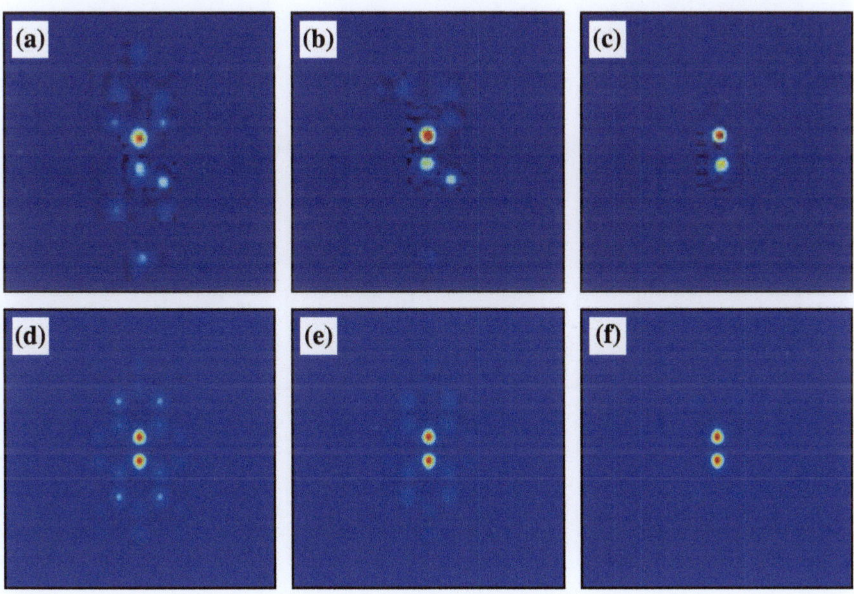

Fig. 5.2 Experimental results (*top*) and numerical simulations (*bottom*) for the formation of a dipole-mode gap soliton in an optically induced triangular lattice. Results are shown for different input powers. **a, d** Discrete diffraction at 10 nW; **b, e** self-focusing at 25 nW; **c, f** dipole-mode gap soliton at 90 nW

gap soliton evolves (Fig. 5.2c). As before, there is a very good agreement between the experimental results and the numerical simulations shown in Fig. 5.2d–f.

The existence of these stable dipole-mode gap solitons already gives an example of nonlinear localization of complex light fields in optically induced photonic lattices. However, the most exciting states in this context are probably the *discrete vortex solitons* as nonlinear localized beams carrying optical vortices, i.e. persistent currents around phase dislocations [16–18]. These particular states will be discussed in more detail in the following sections.

5.1.2 Discrete Vortex Solitons

In general, optical vortices are phase singularities with energy flows rotating in a given direction. At the point of the singularity, the phase is undefined and the intensity vanishes. Therefore, the simplest optical vortex consists of a Gaussian beam with a superimposed helical phase structure resulting in a doughnut-shaped intensity distribution (Fig. 5.3a).

To introduce the common notation, we recall that the electric field $E(\boldsymbol{r}, t)$ can be represented by a complex scalar function $A(\boldsymbol{r})$ describing its slowly varying envelope [cf. (2.14)]. By rewriting this envelope as

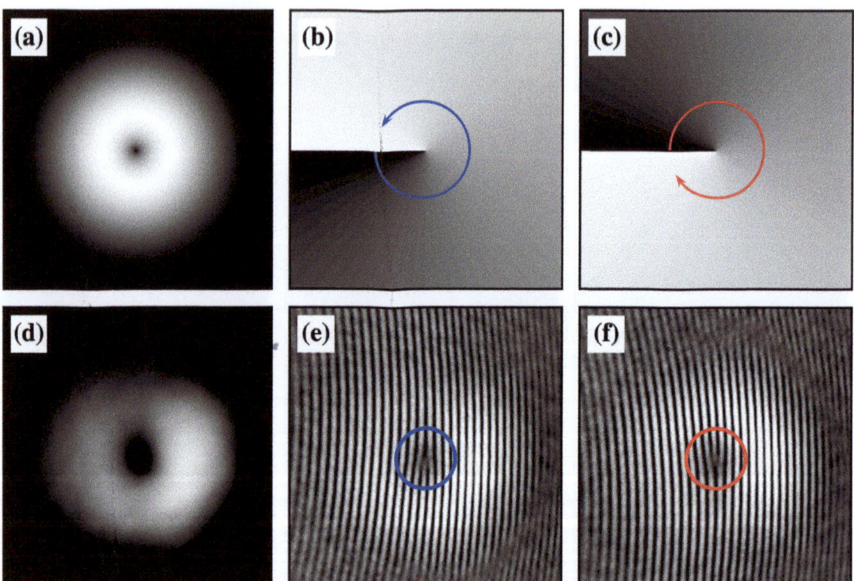

Fig. 5.3 Numerical simulation (*top*) and experimental realization (*bottom*) of a doughnut-shaped vortex beam. **a**, **d** Intensity distribution; **b** phase profile for $M = +1$; **c** phase profile for $M = -1$; **e** interferogram with an inclined broad reference wave for $M = +1$; **f** interferogram with an inclined broad reference wave for $M = -1$; arrows in **b** and **c** indicate the direction of the energy flow; circles in **e** and **f** mark the positions of the phase singularities

$$A(\boldsymbol{r}) = |A(\boldsymbol{r})| \cdot e^{i\phi(\boldsymbol{r})} \tag{5.1}$$

the optical vortex can be characterized by the circulation of the phase gradient around the singularity (Fig. 5.3b, c):

$$M = \frac{1}{2\pi} \oint \nabla \phi d\boldsymbol{S}. \tag{5.2}$$

Here, $d\boldsymbol{S}$ is the element of an arbitrary counter-clockwise closed path around the dislocation. As the phase changes by a multiple of 2π, M is an integer which is commonly denoted as the *topological charge*. Its sign determines the direction of the energy flow around the core (arrows in Fig. 5.3b, c). In case of a positive topological charge, the flow is counter-clockwise while a negative charge corresponds to a clockwise energy flow. For a more detailed discussion of optical vortices and their general properties, we refer to [19].

The experimental analysis of optical vortices is usually done by interfering a vortex beam of topological charge M with a broad plane wave at a small angle. This way, the resulting interference pattern contains characteristic fork dislocations formed by M additional interference fringes as indicated by the circles in Fig. 5.3e, f. These forks in the interference pattern not only allow to determine the position of the phase singularity but also enable a distinction between topological charges of opposite sign by its orientation.

Two-dimensional discrete vortex solitons were first observed experimentally in optically induced photonic lattices in 2004 [20, 21]. In these experiments, a vortex beam of topological charge $M = 1$ was sent into a square photonic lattice optically induced in a positively biased photorefractive crystal (focusing nonlinearity). As before, the lattice was oriented in the diamond-type configuration to minimize the effects of anisotropy. The authors demonstrated that in the nonlinear regime a stable four-lobe intensity distribution is formed with neighboring lobes being $\pi/2$ out of phase, thus preserving the unit topological charge. It is important to note that such nonlinear self-trapped vortex entities in the presence of a focusing nonlinearity do not exist in homogeneous media where vortex beams have been shown to decay into several fundamental solitons that repel and twist around each other during propagation [22].

However, the four-lobe structure of the discrete vortex solitons mentioned above closely resembles the ringlike soliton clusters in homogeneous media discussed in [23]. Therefore, it has been suggested to construct discrete vortex solitons also as a superposition of a finite number of fundamental discrete or gap solitons [24]. Provided that the lattice is deep enough, the positions of the individual solitons (lobes) are fixed by the lattice sites and two adjacent lobes are treated as bell-shaped intensity profiles with homogeneous phases ϕ_i and ϕ_j which are coupled via their evanescent fields only. Depending on the number of lobes, the phase differences $\Phi_{ji} = \phi_j - \phi_i$ are chosen such that the overall phase structure of the soliton cluster contains one or more phase singularities. Under these assumptions, a simple stability criterion can be derived by considering the energy flows between the individual lobes [25, 26].

In order to obtain a stable intensity as well as phase profile, i.e. a (topologically) stable discrete vortex soliton, all the energy flows within the cluster must be balanced. We choose (without loss of generality) a one-dimensional notation with the lobes located at $x = \pm \delta x$. Thus, the total optical amplitude can be written as

$$A(x) = A_1 e^{i\phi_1} e^{-\zeta(x+\delta x)} + A_2 e^{i\phi_2} e^{\zeta(x-\delta x)} \quad \text{for} \quad |x| \ll \delta x. \tag{5.3}$$

The real constants A_1 and A_2 are proportional to the maximum amplitudes of the lobes. The energy flow between the lobes

$$J_x = 2\mathrm{Im}(A^* \partial_x A) = 4\zeta A_1 A_2 e^{-2\zeta \delta x} \sin\left(\phi_2 - \phi_1\right) \tag{5.4}$$

is proportional to the sine of their phase differences. The same result is obtained in two transverse dimensions after integrating along the y axis.

In general, the energy flows are balanced, if the intensities of the lobes do not change during propagation. Therefore, the sum of all intensity flows must vanish for each lobe [24]:

$$\sum_{j=1}^{N} c_{ij} \sin\left(\phi_j - \phi_i\right) \overset{!}{=} 0 \quad , \quad 1 \leq i \leq N. \tag{5.5}$$

Herein, all constants have been collected in the coupling coefficients c_{ij} where i and j denote the respective lobes. In the following sections, we will employ this condition to investigate the stability properties of different types of vortex solitons in optically induced hexagonal lattices.

5.2 Anisotropy-Controlled Stability of Discrete Vortex Solitons

The simplest discrete vortex soliton in hexagonal lattices consists of only three lobes with unit topological charge and has been shown to be stable in isotropic systems [26]. However, as discussed in Sect. 3.3, optically induced hexagonal lattices are strongly affected by the photorefractive anisotropy, resulting in asymmetric coupling between the lattice sites for the original hexagonal lattice wave with $\eta = \sqrt{3}$ and a more symmetric coupling for the stretched lattice with $\eta > \sqrt{3}$. In this section, we demonstrate how the lattice stretching can be employed to tailor the coupling between the lattice sites and to control the stability of three-lobe discrete vortex solitons.

First, we consider propagation in a nonstretched hexagonal lattice. The basic process can easily be understood by analyzing the phase differences and thus the energy flows between the lobes as schematically illustrated in Fig. 5.4.

At the input (Fig. 5.4a), all phase differences are chosen to be equal $\Phi_{ji} = \phi_j - \phi_i = 2\pi/3$. During propagation, the differences in the coupling constants along the diagonal and along the horizontal direction lead to a net transfer of

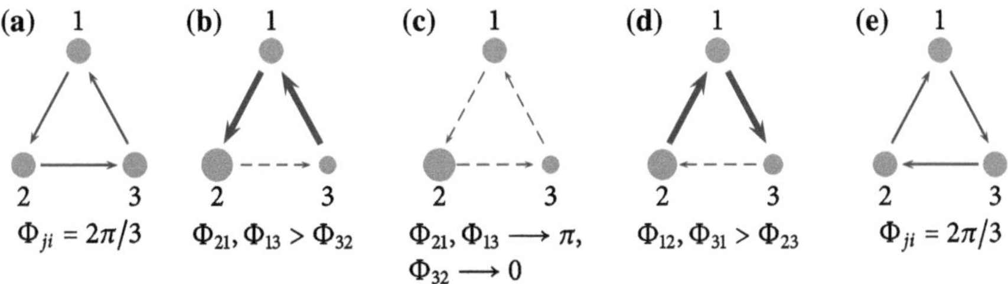

Fig. 5.4 Schematic illustration of the charge flipping process for three-lobe discrete vortex solitons in symmetric hexagonal lattices ($\eta = \sqrt{3}$). **a** Initial state with equal phase differences $\Phi_{ji} = 2\pi/3$; **b** net transfer of energy from lobe 3 to lobe 2 due to asymmetric coupling; **c** vanishing energy transfer for $\Phi_{21}, \Phi_{13} \longrightarrow \pi$ and $\Phi_{32} \longrightarrow 0$; **d** reversed energy transfer from lobe 2 to lobe 3; (**e**) initial intensity distribution with reversed phase structure

energy from lobe 3 to lobe 2 via lobe 1. As a result, the intensity of lobe 2 increases and lobe 3 becomes darker. As the flow condition (5.5) is still fulfilled for lobe 1, its intensity remains almost unchanged (Fig. 5.4b). For a focusing nonlinearity, an increased or decreased intensity of an individual lobe results in an increase or decrease of the propagation constant. Hence, the phase differences Φ_{21} and Φ_{13} grow while Φ_{32} becomes smaller. Due to the sine function in (5.4), this finally leads to vanishing flows between all lobes as $\Phi_{21}, \Phi_{13} \longrightarrow \pi$ and $\Phi_{32} \longrightarrow 0$ (Fig. 5.4c). However, the change of phase differences continues since an intensity difference between the lobes 2 and 3 is still present. Consequently, the energy flows get reversed and the topological charge changes its sign. As shown in Fig. 5.4d, this causes a net energy flow from lobe 2 to lobe 3 and the intensity difference decreases until the original intensity distribution is restored with an inverted phase structure (Fig. 5.4e). The whole process then starts again in a reversed manner until the initial intensity and phase distributions are restored.

The phenomenon described above is called *charge flipping*. It has also been discussed in isotropic systems [24] with symmetric coupling. In this case, the energy flows are balanced and the charge flip has to be triggered by an external perturbation. In contrast, here it arises due to the intrinsic anisotropy of the photorefractive nonlinearity.

The key to stabilize the vortex soliton and prevent the charge flipping is to restore the coupling symmetry of the isotropic system by stretching the lattice wave along the y-direction. This way, the experimental switching between topologically unstable and stable solitons becomes possible.

Figure 5.5 shows the corresponding experimental setup. For the lattice induction, we use the same configuration as in Fig. 3.3, in which a spatial light modulator (SLM1) converts the incoming beam into three interfering plane waves which are imaged onto the front face of a 20 mm long photorefractive SBN crystal which is positively biased with a dc electric field directed along its optical c-axis (focusing nonlinearity). A half wave plate ensures the polarization of the lattice beam to be ordinary such that nonlinear self-action effects can be neglected [27]. A second spatial light modulator (SLM2) is employed to achieve the desired

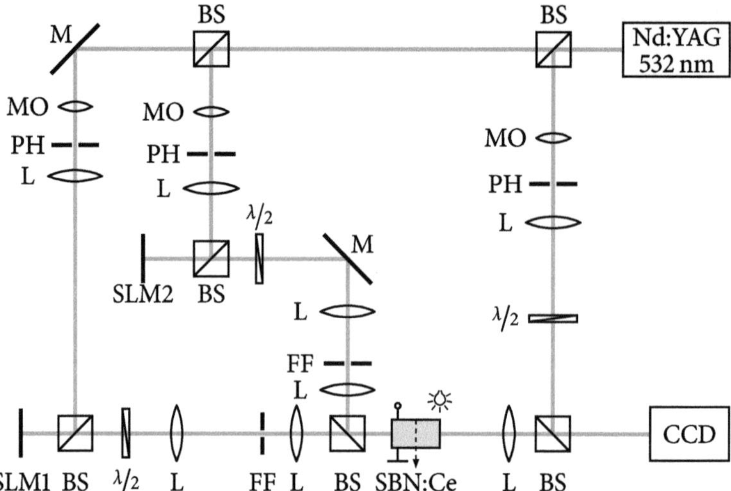

Fig. 5.5 Experimental setup for the observation of discrete vortex solitons. BS: beam splitter; CCD: camera; FF: Fourier filter; L: lens; M: mirror; MO: microscope objective; PH: pinhole; SLM: spatial light modulator

amplitude and phase structure of an incident Gaussian probe beam using the phase imprinting technique.[1] The polarization of the probe beam is extraordinary, so it experiences nonlinearity in its propagation through the crystal. In order to visualize the phase structure of the probe beam, a third beam is derived from the laser. It is passed through a half wave plate to ensure its extraordinary polarization and subsequently sent directly onto the CCD camera to record an interferogram with the probe beam.

Figure 5.6 summarizes the experimental results for the propagation of a three-lobe vortex with unit topological charge in a hexagonal lattice induced by a symmetric lattice wave with a horizontal lattice constant $d_x \approx 30\,\mu$m. The total power of the lattice wave is $P_{\text{latt}} = 50\,\mu$W and the externally applied electric field is approximately 2 kV/cm. Figure 5.6a, d show the corresponding intensity and phase profiles of the input beam. The latter is revealed by the interferogram with the fork-type vortex dislocation (Fig. 5.6d). At low input power ($P_{\text{probe}} \approx 20\,$nW), the output shows discrete diffraction and the phase profile gets distorted. More careful examination reveals that the number of vertical interference fringes coming in and out of the yellow circle in Fig. 5.6b is the same. Therefore, inside the triangle between the three major intensity peaks (initially excited sites), there are two dislocations of opposite charge so that the total topological charge is zero. Furthermore, with an increase of power to $P_{\text{probe}} \approx 150\,$nW, we observe in Fig. 5.6c three well defined intensity spots closely resembling the input structure,

---

[1] A description of the phase imprinting technique to control the amplitude and phase structure of the probe beam is given in appendix C.

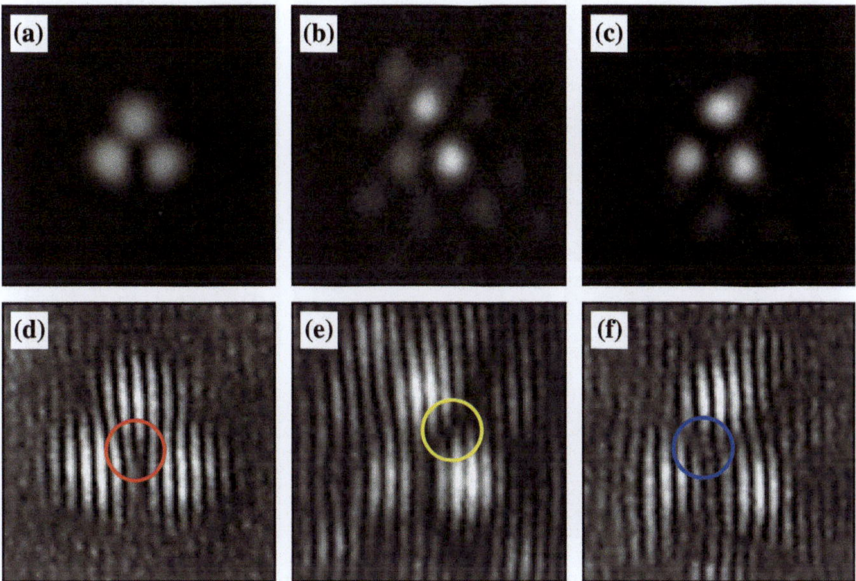

Fig. 5.6 Experimentally obtained intensity (*top*) and phase profiles (*bottom*) for the symmetric hexagonal lattice wave. **a, d** Input vortex beam with unit topological charge (fork up, red circle); **b, e** linear output with zero charge (two opposite dislocations, yellow circle); **c, f** nonlinear ouput with flipped charge (fork down, blue circle)

but the phase profile now shows the opposite topological charge. Thus, in full agreement with our considerations in Fig. 5.4, the nonlinear localization of the intensity is accompanied by the charge flipping effect in experiment.

The experimental results for the stretched lattice with $\eta = 2.5$ are depicted in Fig. 5.7. Obviously, the diffraction in the low power regime is much less pronounced than in the symmetric case and it is hardly visible in Fig. 5.7b. However, the most important result here is that the phase profile is preserved even in the nonlinear output in Fig. 5.7f. Since all the other parameters have been chosen to be the same as in the unstretched lattice in Fig. 5.6, the clear differences in phase profiles can be fully attributed to the stretching of the lattice and the resulting symmetric coupling. Thus, our experiments clearly prove the possibility to control the charge flip in anisotropic hexagonal photonic lattices by stretching the lattice along its vertical direction.

The experimental results for both, the symmetric and the stretched lattice are corroborated by numerical simulations using the anisotropic model and the obtained intensity and phase distributions are shown in Figs. 5.8 and 5.9, respectively. In full agreement with the experiment, the discrete diffraction in the linear regime is clearly visible in the case of the symmetric hexagonal lattice wave (Fig. 5.8b) but is significantly reduced in the stretched lattice (Fig. 5.9b). This effect can be explained by the reduced coupling along the diagonals that suppresses tunneling between the lattice sites.

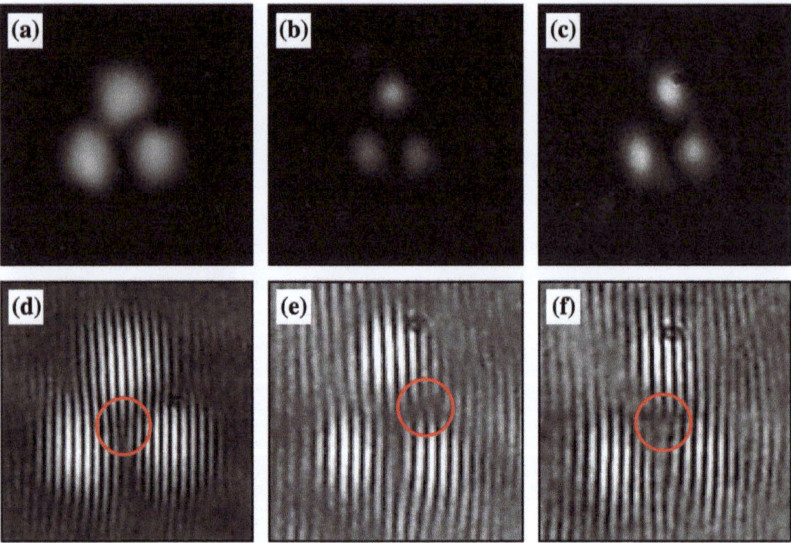

Fig. 5.7 Experimentally obtained intensity (*top*) and phase profiles (*bottom*) for the stretched hexagonal lattice ($\eta = 2.5$). **a, d** Input; **b, e** linear output; **c, f** stable three-lobe discrete vortex soliton; note the same topological charge (fork up, red circle) in all interferograms

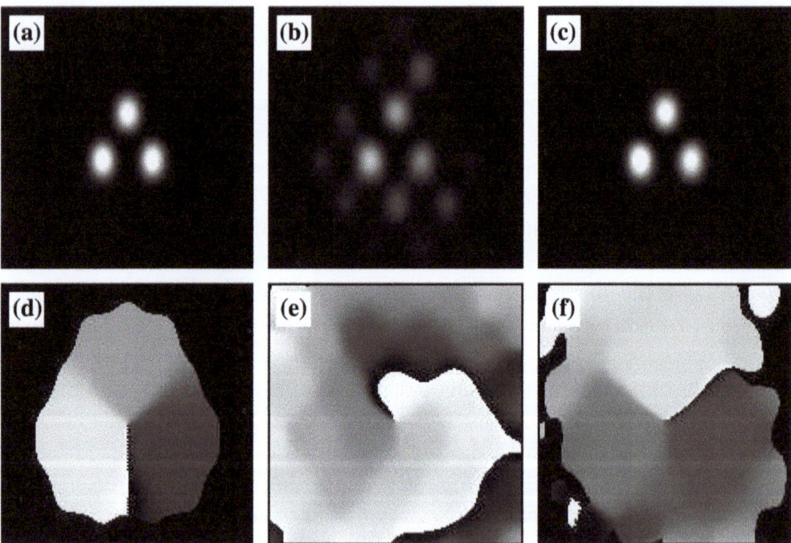

Fig. 5.8 Numerically obtained intensity (*top*) and phase profiles (*bottom*) for the symmetric hexagonal lattice wave ($\eta = \sqrt{3},\ \beta = 5, E_0 = 2.5\,\text{kV/cm}, I_{\text{latt}} = |A_{\text{latt}}|^2 = 2$). **a, d** Input; **b, e** linear output; **c, f** nonlinear output with inverted phase structure; pictures by courtesy of Dr. Tobias Richter [28]

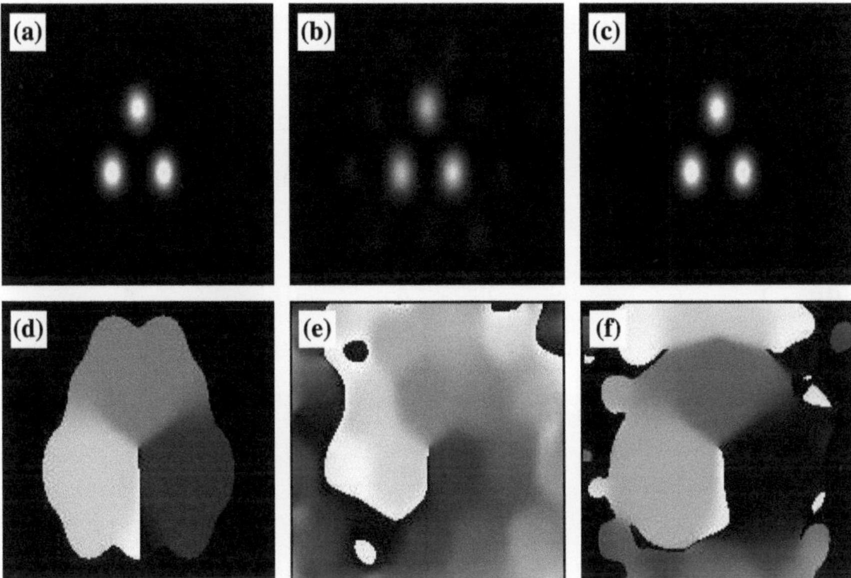

Fig. 5.9 Numerically obtained intensity (*top*) and phase profiles (*bottom*) for the stretched hexagonal lattice ($\eta = 2.5$, $\beta = 5$, $E_0 = 2.5\,\text{kV/cm}$, $I_{\text{latt}} = |A_{\text{latt}}|^2 = 2$). **a, d** Input; **b, e** linear output; **c, f** stable three-lobe discrete vortex soliton; pictures by courtesy of Dr. Tobias Richter [28]

In the nonlinear regime, a localized self-trapped state is formed in both cases (Figs. 5.8c and 5.9c). However, while the phase structures shown in Fig. 5.8d–f reveal the charge flip for the symmetric lattice wave, the discrete vortex soliton in the stretched lattice preserves its topological charge (Fig. 5.9d–f).

In summary, the numerical simulations as well as the experiment clearly show the possibility to control the stability of discrete vortex solitons in optically induced hexagonal lattices by deforming the lattice wave in one transverse direction and thus manipulating the coupling between the lattice sites in order to induce or inhibit the charge flipping effect.

5.3 Double-Charge Discrete Vortex Solitons

In this section, we study ring-shaped clusters consisting of six lobes being arranged in a hexagon with a topological charge of either $M = 1$ or $M = 2$ as illustrated in Fig. 5.10. Following our discussion of the previous section, we choose a stretched hexagonal lattice to counteract the influence of anisotropy.

As before, the stability properties of the ring-shaped clusters can also be understood in terms of energy flows and phase differences between the individual lobes (Fig. 5.10). For the single-charge vortex shown in Fig. 5.10a the initial phase

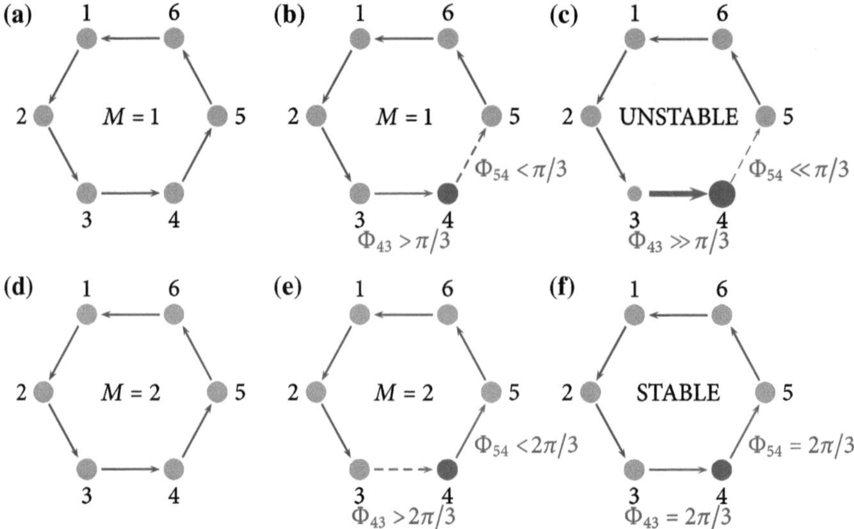

Fig. 5.10 Stability of ring-shaped six-lobe clusters with topological charge $M = 1$ (*top*) and $M = 2$ (*bottom*). **a, d** Input; **b, e** a small perturbation to the initial phase differences causes unbalanced energy flows; **c** for the single-charge vortex, the perturbation grows and vortex becomes unstable; **f** for the double-charge vortex, the growth of the perturbation is inhibited and a stable double-charge discrete vortex soliton becomes possible

difference between two neighboring lobes is $\Phi_{ji} = \pi/3$ and the energy flows are initially balanced due to the lattice stretching. For our stability analysis, we consider a small perturbation to the equality $\Phi_{43} = \Phi_{54} = \pi/3$ and assume (without loss of generality) that $\Phi_{43} > \pi/3$. Since all the other phase differences are fixed, this implies that $\Phi_{54} < \pi/3$, as illustrated in Fig. 5.10b. It is important to recall that the energy flow between two lobes is proportional to the sine of their relative phase difference Φ_{ji} (Fig. 5.11). Hence, for $\Phi_{ji} < \pi/2$, an increasing phase difference results in a stronger energy flow. Consequently, the sum of the flows away from and towards lobe 4 is positive and its intensity grows. For a focusing nonlinearity, this leads to an increase of its propagation constant and Φ_{43} increases as well while Φ_{54} becomes smaller. As a result, the perturbation grows and the single-charge vortex becomes unstable.

The situation changes for the double-charge vortex shown in Fig. 5.10d. In this case, the initial phase differences are $\Phi_{ji} = 2\pi/3$ and thus increasing the phase difference now diminishes the corresponding energy flow (cf. Fig. 5.11). Assuming the same perturbation as before, the sum of the flows away from and towards lobe 4 is now negative and its intensity decreases (Fig. 5.10e). Therefore, the propagation constant decreases as well and Φ_{43} becomes smaller while Φ_{54} becomes larger. This means that the perturbation does not grow in this case and a stable *double-charge discrete vortex soliton* becomes possible (Fig. 5.10f).

The above analysis crucially depends on the sign of the nonlinearity. So far, we have assumed a focusing nonlinearity and thus an increase of the propagation

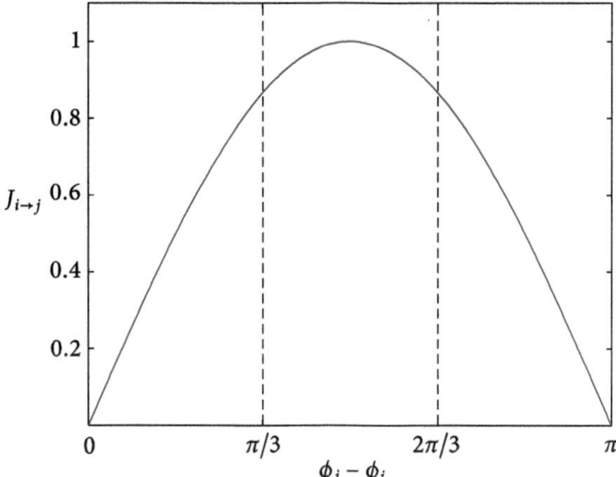

Fig. 5.11 Intensity flow $J_{i \to j}$ between two neighboring lobes (normalized to 1) as a function of the relative phase difference $\Phi_{ji} = \phi_j - \phi_i$.

constant with an increasing intensity. However, a defocusing nonlinearity leads to the opposite situation in which an increased intensity lowers the propagation constant. In this case, the same argumentation as above gives exactly inverted stability properties, resulting in the fact that single-charge vortices become stable while their double-charge counterparts are unstable.

On the one hand, the result of this simple consideration that raising the topological charge can increase the stability is particularly surprising and has no counterpart in homogeneous media where higher-charge discrete vortices are typically unstable [16]. On the other hand, it fully agrees with earlier theoretical works in which the isotropic model has been used to study the stability properties of vortex solitons in hexagonal [29] as well as modulated Bessel photonic lattices [30].

5.3.1 Focusing Nonlinearity

We start our investigations with the case of a focusing nonlinearity and use the same experimental setup as before (Fig. 5.5). The two phase modulators are employed to induce a stretched hexagonal lattice ($d_x = 27\,\mu$m, $\eta = 2.5$) and to generate single- and double-charge vortices, respectively. The characteristics of the beams are otherwise identical, and thus any differences in the dynamics are solely due to the different input phases. The total power of the lattice wave is $P_{\text{latt}} = 75\,\mu$W and the externally applied electric field is approximately 2.2 kV/cm. We selectively vary the probe beam input power to effectively move from the linear regime (low power, $P_{\text{probe}} \approx 50\,$nW) to the nonlinear regime (high power, $P_{\text{probe}} \approx 550\,$nW). The single-charge vortex input is shown in Fig. 5.12a. Its intensity distribution has a form of a necklace with six intensity peaks whose positions correspond to the lattice sites (index maxima). At low input power, the beam undergoes discrete diffraction and a complete loss of the initial six site input

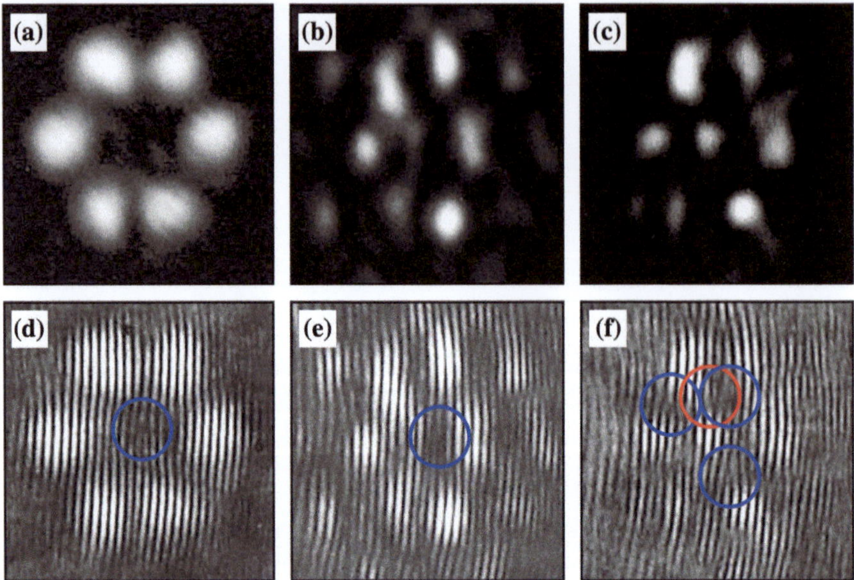

Fig. 5.12 Experimentally obtained intensity (*top*) and phase profiles (*bottom*) for the single-charge vortex. Circles indicate positions of vortices with topological charge $M = +1$ (*blue*) or $M = -1$ (*red*). **a, d** Intensity and phase distribution of the input single-charge vortex beam; **b, e** beam profile and phase at the output for low input power; **c, f** output for high input power

state (Fig. 5.12b). At high power, the initial six site intensity profile also changes significantly after propagation (Fig. 5.12c), showing strong intensity modulations and even filling in the central lattice site. Furthermore, in the phase profile multiple vortices are seen to appear, further indicating a breakdown of the single-charge state (Fig. 5.12f).

In the case of the double-charge vortex input (Fig. 5.13), we again observe discrete diffraction with low input power (Fig. 5.13b). However, the result changes dramatically when the power is increased (Fig. 5.13c). We observe that now the six-site input structure is preserved in the nonlinear propagation. Interestingly, while the overall phase winding is still 4π, it is seen that the initial double-charge singularity has split into two single-charge vortices (Fig. 5.13f). This splitting of the higher-order singularity has been shown to be due to an inherent topological instability in the higher phase-winding [31]. The topological breakdown in the linear (low power) part of the field further indicates that the stability of the 4π phase winding across the six sites is due to the interplay of the nonlinearity and the local phase of the high-power sites suppressing the development of a dynamical instability [29]. However, we find that this stability is critically dependent on the symmetry of the lattice, with a decrease in the lattice stretching (and thus a corresponding decrease in the symmetry of the underlying modulated refractive index) leading to a dynamical instability in the double-charge state as well.

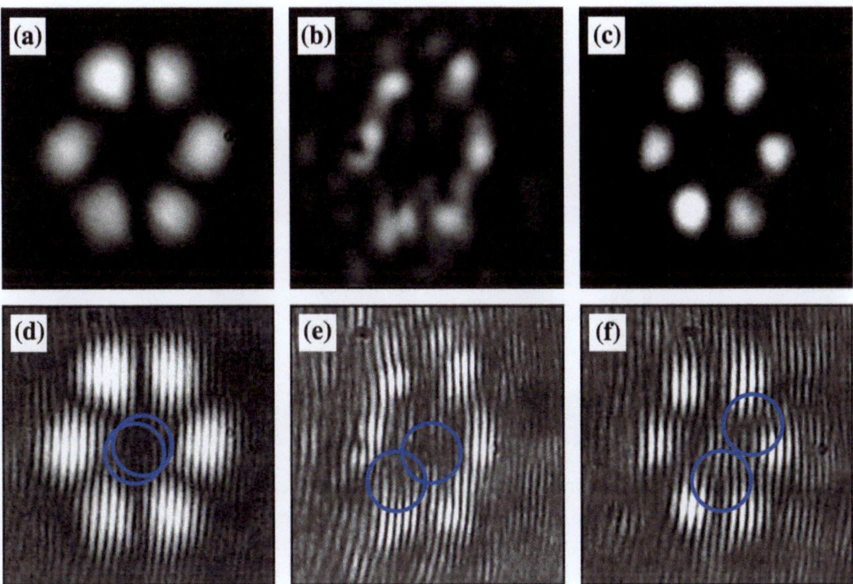

Fig. 5.13 Experimentally obtained intensity (*top*) and phase profiles (*bottom*) for the double-charge vortex. Blue circles indicate positions of vortices with topological charge $M = +1$. **a, d** Intensity and phase distribution of the input double-charge vortex beam; **b, e** beam profile and phase at the output for low input power; **c, f** double-charge discrete vortex soliton

The phase interferogram in Fig. 5.13f also indicates an additional pair of single-charge vortices with opposite charges inside the vortex structure (not marked by circles). However, this additional pair does not affect the stability of the 4π phase winding, and it can be fully attributed to inevitable experimental noise in this region of low intensity.

In numerical simulations, we first consider the case of a six-site initial state with a single-charge vortex phase of the form shown in Fig. 5.14a, e with either low or high power propagating a distance of $z = 20\,\mathrm{mm}$ in the lattice. For the low input power case (Fig. 5.14b, f) we see that, as in the experiment, the vortex beam undergoes strong diffraction and break-up. If instead a high input power is considered (Fig. 5.14c, g), the vortex maintains much of its form. Some intensity fluctuations are evident, and more importantly, the vortex phase has deteriorated showing breakdown of the initial single-charge vortex circulation. It must be noted that the break-up is clearly less than that observed in the experiment and this discrepancy is attributed to the higher anisotropy of the experimental lattice leading to a larger instability growth rate. The strong instability becomes evident for longer propagation distances as shown in Fig. 5.14d, h for $z = 280\,\mathrm{mm}$.

In Fig. 5.15, we consider the same input beam intensities but change the phase to that of a double-charge vortex, as shown in Fig. 5.15e. The low power output in Fig. 5.15b, f appears similar to the single-charge case, exhibiting diffraction and break-up of the vortex. In contrast, the high power output in Fig. 5.15c, d appears

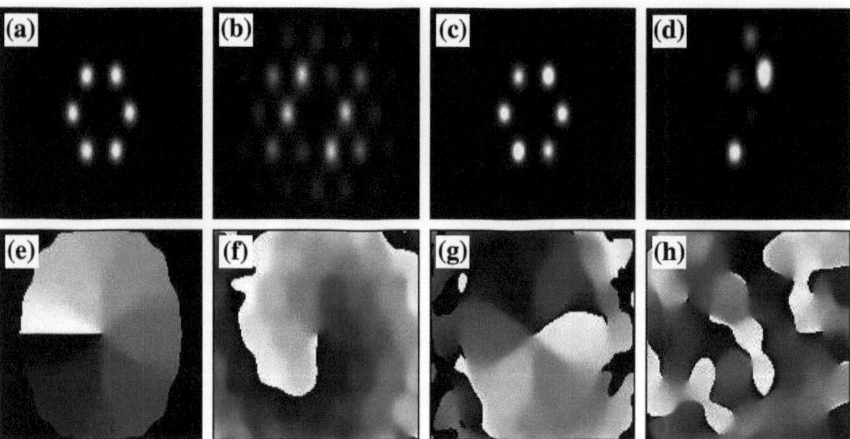

Fig. 5.14 Numerical simulation of the intensity (*top*) and phase profiles (*bottom*) for the single-charge vortex ($E_0 = 2.5\,\text{kV/cm}$, $I_{\text{latt}} = |A_{\text{latt}}|^2 = 1$, $\beta = 3$). **a, e** Input; **b, f** beam profile at $z = 20\,\text{mm}$ for low input power; **c, g** high power output at $z = 20\,\text{mm}$; **d, h** high power output at $z = 280\,\text{mm}$; pictures by courtesy of Dr. Tobias Richter [32]

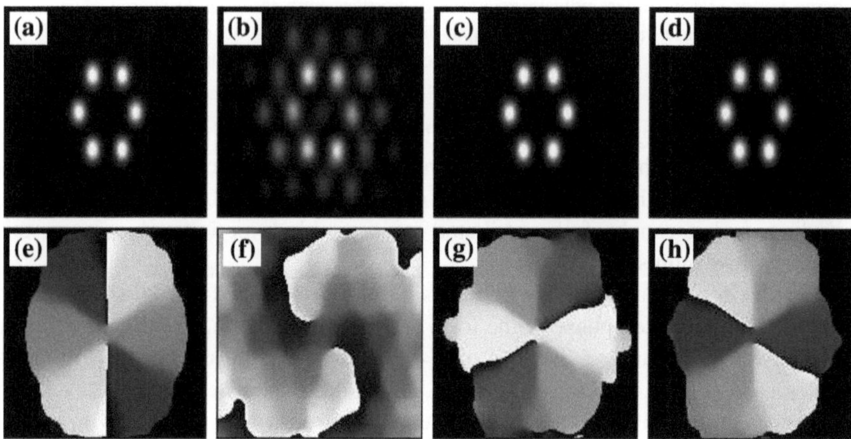

Fig. 5.15 Numerical simulation of the intensity (*top*) and phase profiles (*bottom*) for the double-charge vortex ($E_0 = 2.5\,\text{kV/cm}$, $I_{\text{latt}} = |A_{\text{latt}}|^2 = 1$, $\beta = 3$). **a, e** Input; **b, f** beam profile at $z = 20\,\text{mm}$ for low input power; **c, g** double-charge discrete vortex soliton at $z = 20\,\text{mm}$; **d, h** double charge discrete vortex soliton at $z = 280\,\text{mm}$; pictures by courtesy of Dr. Tobias Richter [32]

unchanged in the intensity profile with a well-pronounced double-charge vortex phase (Fig. 5.15g, h). Similar to the experimental results, the separation of the double-charge phase singularity into two single-charge singularities is observed. However, the phase circulation around a contour tracing the six high intensity sites is well defined and equals 4π.

5.3.2 Defocusing Nonlinearity

As discussed above, the stability properties of single- and double-charge vortices with focusing nonlinearity are expected to be inverted when the nonlinearity is changed from focusing to defocusing. Similar to the focusing case, numerical simulations are performed using the anisotropic model, but with a reversed sign of the nonlinearity by using $E_0 = -2.5\,$kV/cm.

Figure 5.16 summarizes the results for the single-charge vortex and clearly demonstrates the inverted stability properties caused by the defocusing nonlinearity. In contrast to the focusing case (Fig. 5.14), intensity and phase profile of the input structure are preserved and a stable single-charge vortex soliton is formed. It should be noted however, that in the low intensity regime (Fig. 5.16b) the diffraction is much less pronounced than in the presence of a focusing nonlinearity and hardly visible for propagation distances of 20 mm. The same result is obtained for the low intensity double-charge vortex shown in Fig. 5.17b. Moreover, compared to the focusing case, the instability is weaker and more evident in the phase than in the intensity (Figs. 5.17g, h). Overall, the numerical simulations well confirm the theoretical prediction of inverted stability properties in the defocusing case resulting in a stable single-charge vortex soliton and an unstable double-charge vortex.

Experimentally, the nonlinearity is made defocusing by simply inverting the external bias voltage. We consider a bias field of $E_0 \approx 1.6\,$kV/cm antiparallel to the optical axis. The total power of the lattice beam is $P_{\text{latt}} \approx 50\,\mu$W and we use a stretched lattice with the same lattice constants as in the focusing case. Notice that

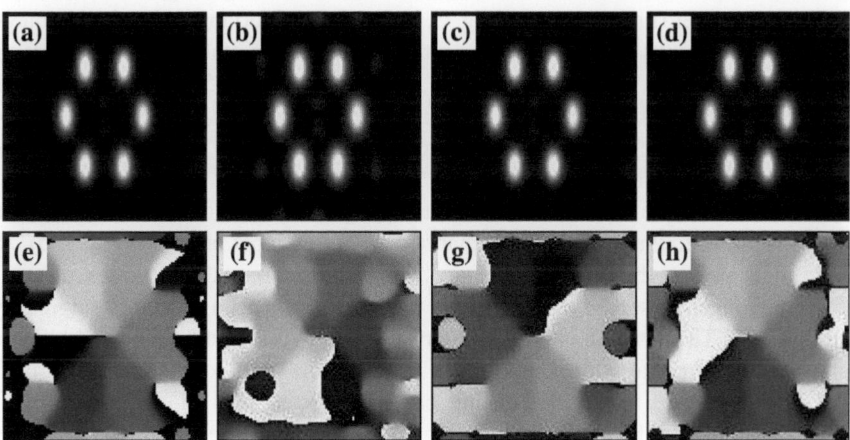

Fig. 5.16 Numerical simulation of the intensity (*top*) and phase profiles (*bottom*) for the single-charge vortex in the presence of a defocusing nonlinearity ($E_0 = -2.5\,$kV/cm, $I_{\text{latt}} = |A_{\text{latt}}|^2 = 4$, $\beta = 2$). **a, e** Input; **b, f** beam profile at $z = 20\,$mm for low input power; **c, g** high power output at $z = 20\,$mm; **d, h** high power output at $z = 280\,$mm; pictures by courtesy of Dr. Tobias Richter [32]

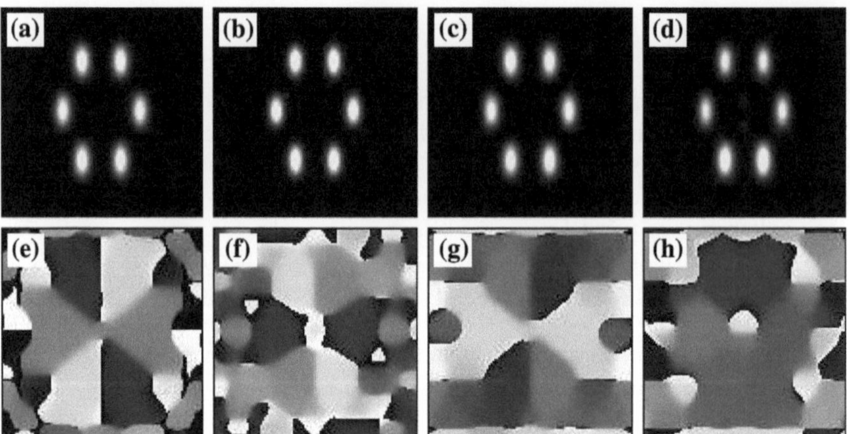

Fig. 5.17 Numerical simulation of the intensity (*top*) and phase profiles (*bottom*) for the double-charge vortex in the presence of a defocusing nonlinearity ($E_0 = -2.5\,\text{kV/cm}$, $I_{\text{latt}} = |A_{\text{latt}}|^2 = 4$, $\beta = 2$). **a, e** Input; **b, f** beam profile at $z = 20\,\text{mm}$ for low input power; **c, g** high power output at $z = 20\,\text{mm}$; **d, h** high power output at $z = 280\,\text{mm}$; pictures by courtesy of Dr. Tobias Richter [32]

now the lattice acquires a honeycomb structure, i.e. light intensity maxima of the lattice forming beams lead to minima of the corresponding refractive index pattern. For each vortex input, we consider two different input beam powers, low power ($I_{\text{probe}} \approx 30\,\text{nW}$) and high power ($I_{\text{probe}} \approx 160\,\text{nW}$).

First, we consider the single-charge vortex input shown in Fig. 5.18a, d. In good agreement with the numerical simulations, we see only very weak diffraction in the low power (linear) regime (Fig. 5.18b). More importantly, the single-charge vortex phase breaks up and we observe the emergence of other vortices indicating that the input beam profile is not stable at low power (Fig. 5.18e). In contrast, at high power, we find that both the intensity (Fig. 5.18c) and phase profile (Fig. 5.18f) are well preserved, in strong contrast to the observations in the presence of a focusing nonlinearity.

In the case of a double-charge vortex (Fig. 5.19), we observe a diffraction pattern similar to that in the single-charge case at low input power (Fig. 5.19b). At high input power, the output still shows some diffraction and the vortex phase is again no longer preserved as we are able to identify only a single vortex singularity (Fig. 5.19f). This is again in strong contrast to the focusing case. We would like to stress again here that while the stability of the single- and double-charge vortices has been swapped in the defocusing case, the appearance of the instability is somewhat different between the focusing and defocusing cases. In the former case, we observed strong intensity modulations which made clear that the single-charge vortex is unstable. In the defocusing case, the instability development appears to be weaker and to be more evident in the phase than in the intensity. However, we can still conclude that the stability properties of the vortices in the defocusing case are inverse to those in the focusing case.

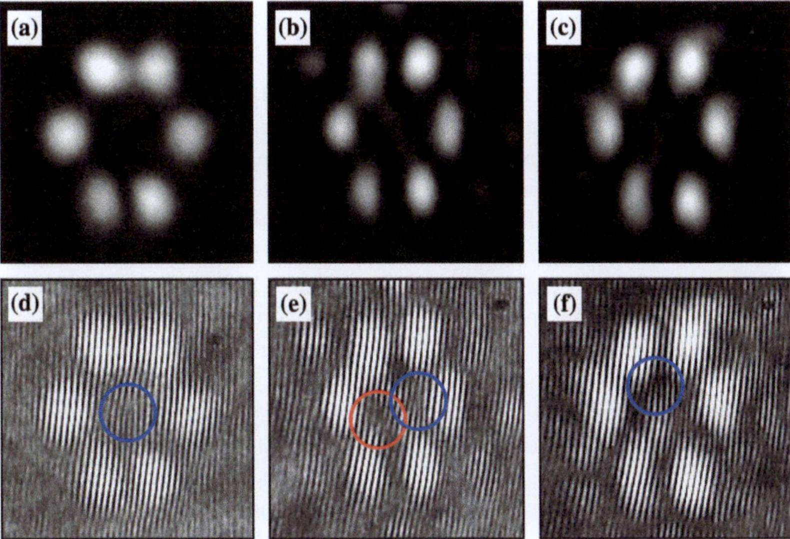

Fig. 5.18 Experimentally obtained intensity (*top*) and phase profiles (*bottom*) for the single-charge vortex in the presence of a defocusing nonlinearity. Circles indicate positions of vortices with topological charge $M = +1$ (*blue*) or $M = -1$ (*red*). **a, d** Intensity and phase distribution of the input single-charge vortex beam; **b, e** beam profile and phase at the output for low input power; **c, f** beam profile and phase at the output for high input power

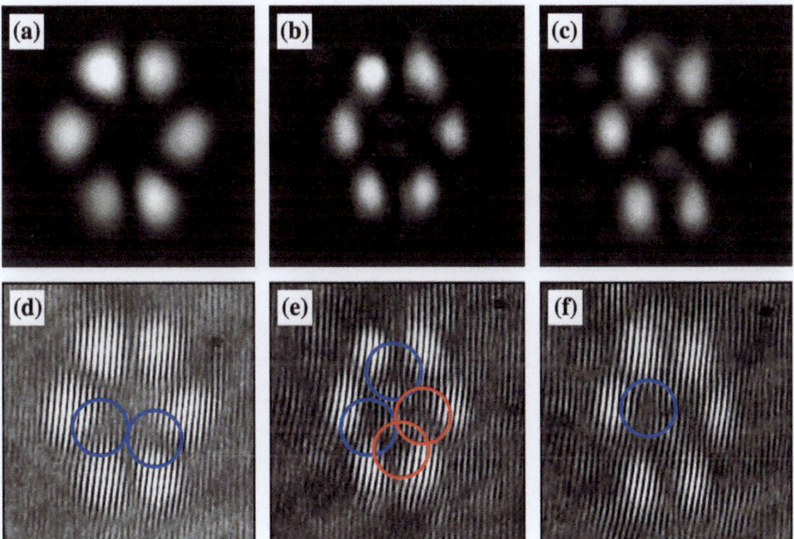

Fig. 5.19 Experimentally obtained intensity (*top*) and phase profiles (*bottom*) for the double-charge vortex in the presence of a defocusing nonlinearity. Circles indicate positions of vortices with topological charge $M = +1$ (*blue*) or $M = -1$ (*red*). **a, d** Intensity and phase distribution of the input single-charge vortex beam; **b, e** beam profile and phase at the output for low input power; **c, f** beam profile and phase at the output for high input power

5.4 Multivortex Solitons

So far, the phase profiles of the vortex solitons investigated in this thesis contained only one isolated phase singularity. Here, we extend this concept and consider structures with multiple phase singularities. In general, multivortex states appear naturally in systems with repulsive inter-particle interactions where they can be confined by external potentials. For attractive interaction, multivortex structures are known to be unstable and have only been observed as infinite periodic waves [33]. However, it was recently predicted theoretically [26] that hexagonal photonic lattices can support stable spatially localized multivortex states. Based on isotropic numerical simulations, this work showed that localized states with high vorticity are stable whereas their counterparts with only one phase singularity experience strong topological instabilities [34].

As an example of such a localized multivortex structure, we study clusters of seven lobes being arranged in hexagonal shape as shown in Fig. 5.20. As before, the energy flows between the lobes determine the topological stability. Due to the hexagonal shape, the outer lobes of the cluster have three neighbors, whereas the inner one has six. Therefore, a balanced energy flow within the cluster can only be achieved if the flows between the center lobe and each outer one are exactly twice as high as the flows between the outer lobes themselves (cf. arrows in Fig. 5.20). Under this constraint, the stability criterion (5.5) can only be fulfilled for symmetric coupling, i.e. all c_{ij} are equal. As a result, the above structure is expected to be stable only in the stretched lattice while the unbalanced energy flows should cause instabilities in case of a symmetric hexagonal lattice wave. In fact, a complex phase dynamics including charge flips of the cluster has been demonstrated [35]. Here, we focus on the observation of stable multivortex solitons in stretched hexagonal lattices and stress that this observation provides the first evidence (in any field of physics) of stable multivortex clusters in systems with attractive nonlinear interaction.

Our experimental setup is shown schematically in Fig. 5.21. Similar to the optical induction setup introduced in Fig. 3.2, a laser beam is split into four separate beams of equal power ($\approx 0.1\,\mathrm{mW}$) by passing through a configuration of two Mach-Zehnder-type interferometers (bottom part in Fig. 5.21). One of the beams is blocked while the other three interfere inside a 15 mm long

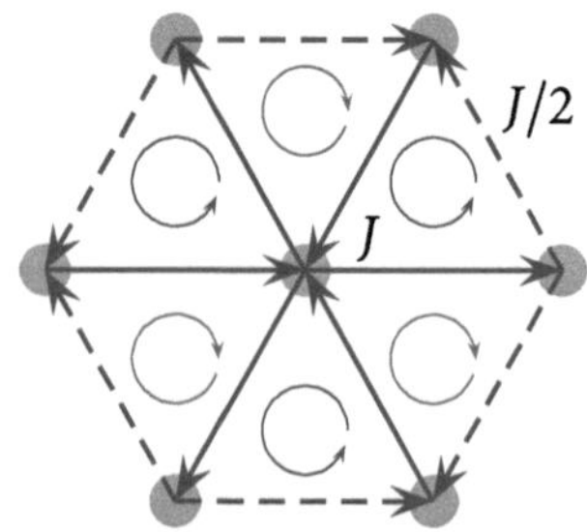

Fig. 5.20 Schematic illustration of the energy flows within a hexagonal seven-lobe cluster. Vortices are indicated by blue arrows for topological charge $M = +1$ and red arrows for $M = -1$

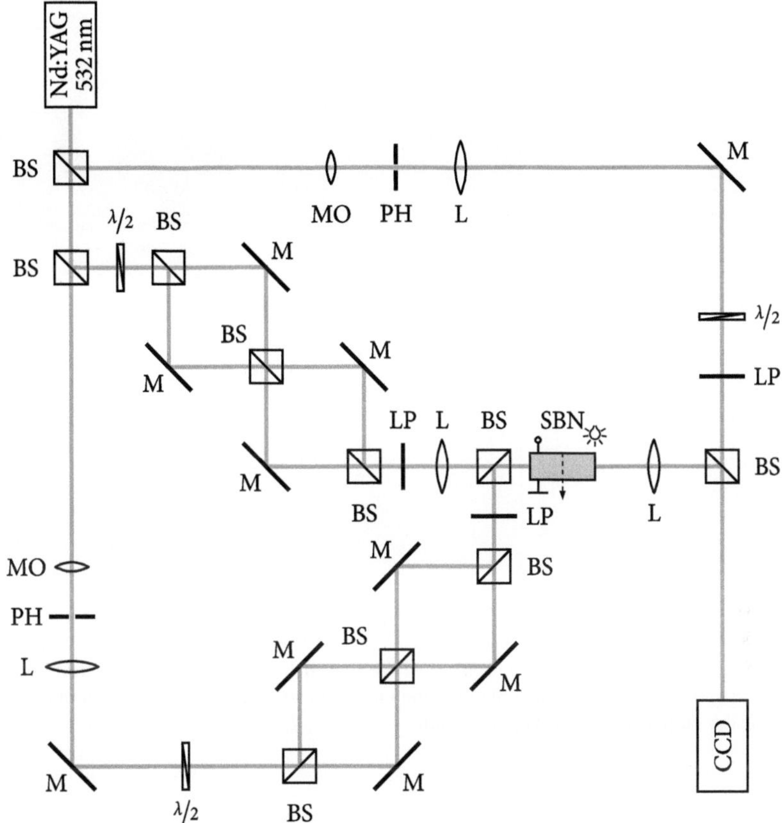

Fig. 5.21 Experimental setup for the observation of multivortex solitons. *BS*: beam splitter; *CCD*: camera; *L*: lens; *LP*: linear polarizer; *M*: mirror; *MO*: microscope objective; *PH*: pinhole

photorefractive SBN crystal which is positively biased with a dc field of $E_0 \approx$ 2 kV/cm, thus providing a focusing nonlinearity. The angles of the interfering beams are adjusted to give a stretched hexagonal lattice wave with $\eta = 2.5$ and a horizontal lattice constant of $d_x = 27\,\mu$m. A polarizer ensures the ordinary polarization necessary for linear propagation through the crystal. The output of the crystal is then imaged onto a CCD camera.

As we have already seen in Fig. 3.7, the hexagonal lattice wave naturally contains optical vortices nested between the lattice sites with neighboring vortices having opposite topological charges $M = +1$ and $M = -1$, respectively. Therefore, in order to generate the multivortex probe beam of hexagonal symmetry (cf. Fig. 5.20), another beam is passed through an identical arrangement of interferometers (top part of Fig. 5.21) and the resulting three beams are subsequently focused onto the front face of the crystal. In contrast to the lattice wave, the polarization is made extraordinary in this case to enable nonlinear propagation. The angles are adjusted such that the probe beam has the same symmetry as the

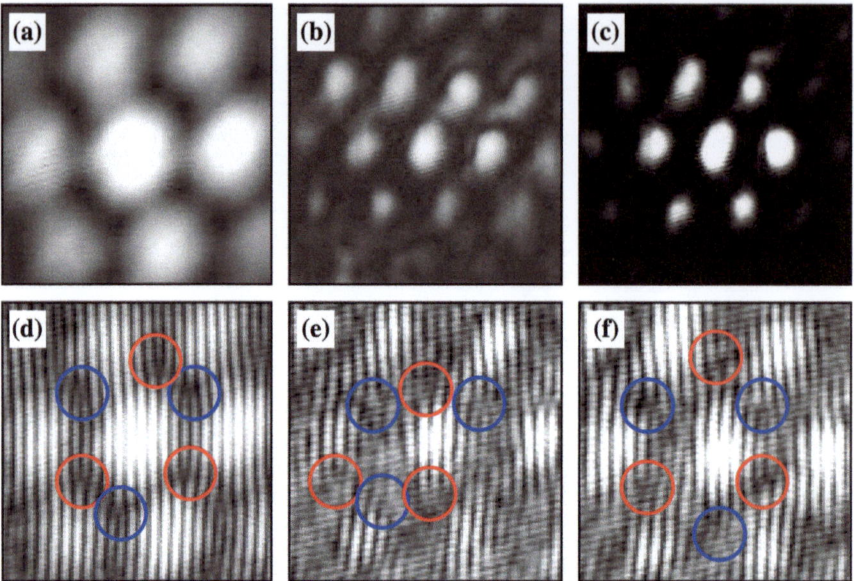

Fig. 5.22 Experimentally obtained intensity (*top*) and phase profiles (*bottom*) for a multivortex soliton in a stretched hexagonal lattice $(\eta = 2.5)$. Circles indicate positions of vortices with topological charge $M = +1$ (*blue*) or $M = -1$ (*red*). **a, d** Intensity and phase distribution of the input beam; **b, e** beam profile and phase at the output for low input power; **c, f** beam profile and phase at the output for a multivortex soliton

induced lattice. In order to achieve localization of the probe beam, the spatial frequency spectrum of the constituent beams is kept much broader than that of the lattice forming beams, resulting in an input probe beam having the form of seven distinctive spots forming a hexagonal pattern with the same period as the lattice and containing six vortices as shown in Fig. 5.22a, d. The vortex positions, indicated by blue and red circles in Fig. 5.22d are again visualized by interfering the probe beam with an inclined broad reference beam which is sent directly onto the camera. At low input power $(P_{\text{probe}} \approx 20\,\text{nW})$, the diffraction of the probe beam leads to the broad output distribution shown in Fig. 5.22b. However, at high power $(P_{\text{probe}} \approx 1\,\mu\text{W})$, the structure becomes localized and the output intensity distribution features seven well-pronounced spots closely resembling the input (Fig. 5.22c). The corresponding phase interferograms of the reference beam and the probe beam at low and high powers are shown in Fig. 5.22e, f, respectively. It is clearly visible that at low power (in the linear regime) the initial phase profile becomes distorted. The six initial vortices can still be found in the output field but their positions are changed. In contrast, for high input power of the probe beam (in the nonlinear regime) not only the beam intensity becomes self-trapped, but also the phase profile retains exactly the same hexagonal vortex pattern of the input beam (Fig. 5.22f).

We stress again that the observation of stable multivortex solitons requires stretching of the hexagonal lattice. When the lattice is exactly of hexagonal shape, self-trapping in the form of seven intensity spots can still be observed; however, the phase distribution becomes random, and it does not contain a regular cluster of vortices similar to that shown in Fig. 5.21. Moreover, in this case the output profile experiences strong deformations even for slight perturbations of the input beam. In a sharp contrast, the multivortex solitons in the stretched lattice are remarkably robust and are basically insensitive to small deformations of the input beam. For the sake of completeness, our experimental observations are again accompanied by

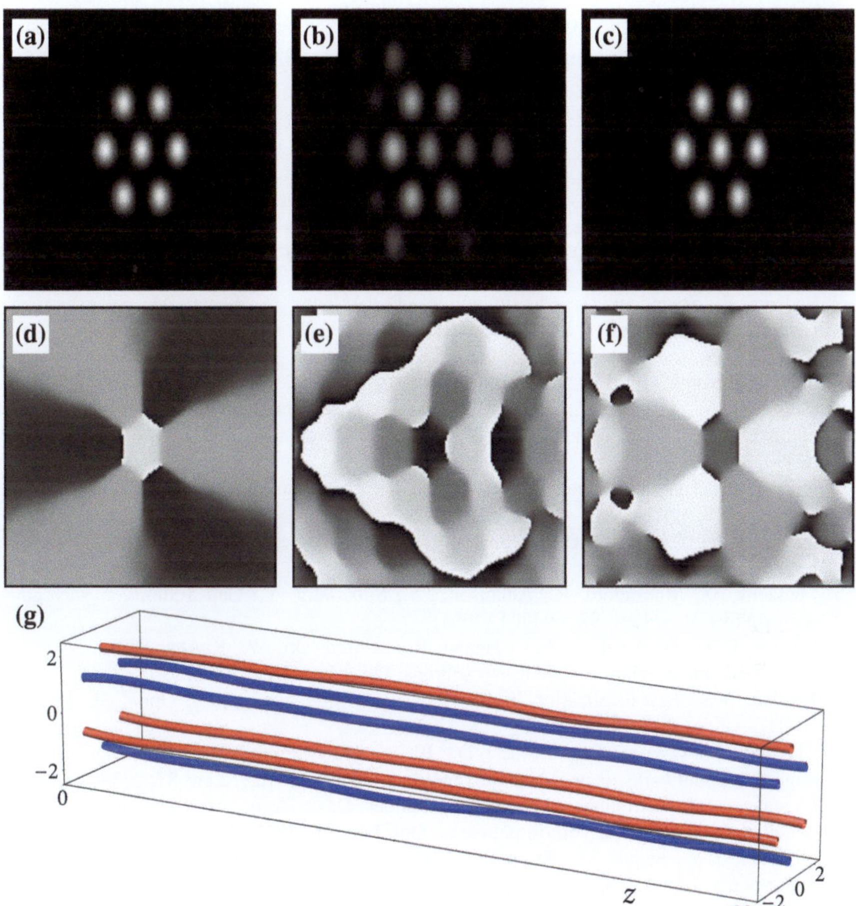

Fig. 5.23 Numerically obtained intensity (*first row of panels*), phase profiles (*second row of panels*) and vortex trajectories (*bottom*) for a multivortex soliton in the stretched hexagonal lattice ($\eta = 2.5, \beta = 3, E_0 = 2.5\,\mathrm{kV/cm}, I_{\mathrm{latt}} = |A_{\mathrm{latt}}|^2 = 1$). **a, d** Input; **b, e** linear output; **c, f** multivortex soliton; **g** 3D vortex trajectories during propagation; pictures by courtesy of Dr. Tobias Richter [35]

numerical simulations using the anisotropic model. The results summarized in Fig. 5.23 clearly confirm our experimental observations. At low power, the seven-lobe input shows diffraction leading to a distortion of the phase structure. However, in agreement with the experiment (cf. Fig. 5.22e), the initial six vortices can still be located between the original seven spots as shown in Fig. 5.23e. In the nonlinear regime, the output intensity distribution shows seven well defined spots (Fig. 5.23c), and we observe a very regular phase profile with all six vortices preserved intact (Fig. 5.23f). The surface plot in Fig. 5.23g shows weak oscillations of the vortex positions which we ascribe to internal oscillatory modes of the soliton. An important conclusion drawn from these simulations is that even for propagation distances much longer than the crystal length, these modes do not grow. This indicates the stability of the multivortex soliton against small perturbations as it has been observed in experiment as well.

References

1. Christodoulides, D.N., Joseph, R.I.: Discrete self-focusing in nonlinear arrays of coupled waveguides. Opt. Lett. **13**, 794 (1988)
2. Lederer, F., Stegeman, G., Christodoulides, D., Assanto, G., Segev, M., Silberberg, Y.: Discrete solitons in optics. Phys. Rep. **463**, 1 (2008)
3. Neshev, D.N., Ostrovskaya, E., Kivshar, Y.S., Krolikowski, W.: Spatial solitons in optically induced gratings. Opt. Lett. **28**, 710 (2003)
4. Fleischer, J.W., Carmon, T., Segev, M., Efremidis, N.K., Christodoulides, D.N.: Observation of discrete solitons in optically induced real time waveguide arrays. Phys. Rev. Lett. **90**, 023902 (2003)
5. Feng, J.: Alternative scheme for studying gap solitons in an infinite periodic Kerr medium. Opt. Lett. **18**, 1302 (1993)
6. Nabiev, R.F., Yeh, P., Botez, D.: Spatial gap solitons in periodic nonlinear structures. Opt. Lett. **18**, 1612 (1993)
7. Mandelik, D., Morandotti, R., Aitchison, J.S., Silberberg, Y.: Gap Solitons in waveguide arrays. Phys. Rev. Lett. **92**, 093904 (2004)
8. Neshev, D.N., Sukhorukov, A.A., Hanna, B., Krolikowski, W., Kivshar, Y.S.: Controlled generation and steering of spatial gap solitons. Phys. Rev. Lett. **93**, 083905 (2004)
9. Kivshar, Y.S.: Self-localization in arrays of defocusing waveguides. Opt. Lett. **18**, 1147 (1993)
10. Christodoulides, D.N., Eugenieva, E.D.: Blocking and routing discrete solitons in two-dimensional networks of nonlinear waveguide arrays. Phys. Rev. Lett. **87**, 233901 (2001)
11. Eugenieva, E.D., Efremidis, N.K., Christodoulides, D.N.: Design of switching junctions for two-dimensional discrete soliton networks. Opt. Lett. **26**, 1978 (2001)
12. Fleischer, J.W., Segev, M., Efremidis, N.K., Christodoulides, D.N.: Observation of two-dimensional discrete solitons in optically induced nonlinear photonic lattices. Nature **422**, 147 (2003)
13. Terhalle, B., Desyatnikov, A.S., Bersch, C., Träger, D., Tang, L., Imbrock, J., Kivshar, Y.S., Denz, C.: Anisotropic photonic lattices and discrete solitons in photorefractive media. Appl. Phys. B **86**, 399 (2007)
14. Rosberg, C.R., Neshev, D.N., Sukhorukov, A.A., Krolikowski, W., Kivshar, Y.S.: Observation of nonlinear self-trapping in triangular photonic lattices. Opt. Lett. **32**, 397 (2007)

15. Rose, P., Richter, T., Terhalle, B., Imbrock, J., Kaiser, F., Denz, C.: Discrete and dipole-mode gap solitons in higher-order nonlinear photonic lattices. Appl. Phys. B **89**, 521 (2007)
16. Desyatnikov, A.S., Torner, L., Kivshar, Y.S.: Optical vortices and vortex solitons. Prog. Opt. **47**, 219 (2005)
17. Malomed, B.A., Kevrekidis, P.G.: Discrete vortex solitons. Phys. Rev. E **64**, 026601 (2001)
18. Yang, J., Musslimani, Z.H.: Fundamental and vortex solitons in a two-dimensional optical lattice. Opt. Lett. **28**, 2094 (2003)
19. Basistiy, I.V., Soskin, M.S., Vasnetsov, M.V.: Optical wavefront dislocations and their properties. Opt. Commun. **119**, 604 (1995)
20. Neshev, D.N., Alexander, T.J., Ostrovskaya, E.A., Kivshar, Y.S., Martin, H., Makasyuk, I., Chen, Z.: Observation of discrete vortex solitons in optically induced photonic lattices. Phys. Rev. Lett. **92**, 123903 (2004)
21. Fleischer, J.W., Bartal, G., Cohen, O., Manela, O., Segev, M., Hudock, J., Christodoulides, D.N.: Observation of vortex-ring "discrete" solitons in 2D photonic lattices. Phys. Rev. Lett. **92**, 123904 (2004)
22. Firth, W.J., Skryabin, D.V.: Optical solitons carrying orbital angular momentum. Phys. Rev. Lett. **79**, 2450 (1997)
23. Desyatnikov, A.S., Kivshar, Y.S.: Rotating optical soliton clusters. Phys. Rev. Lett. **88**, 053901 (2002)
24. Alexander, T.J., Sukhorukov, A.A., Kivshar, Y.S.: Asymmetric vortex solitons in nonlinear periodic lattices. Phys. Rev. Lett. **93**, 063901 (2004)
25. Richter, T.: Stability of anisotropic gap solitons in photorefractive media. Ph.D. thesis, TU Darmstadt (2008)
26. Alexander, T.J., Desyatnikov, A.S., Kivshar, Y.S.: Multivortex solitons in triangular photonic lattices. Opt. Lett. **32**, 1293 (2007)
27. Efremidis, N.K., Sears, S., Christodoulides, D.N., Fleischer, J.W., Segev, M.: Discrete solitons in photorefractive optically induced photonic lattices. Phys. Rev. E **66**, 046602 (2002)
28. Terhalle, B., Göries, D., Richter, T., Rose, P., Desyatnikov, A.S., Kaiser, F., Denz, C.: Anisotropy-controlled topological stability of discrete vortex solitons in optically induced photonic lattices. Opt. Lett. **35**, 604 (2010)
29. Law, K.J.H., Kevrekidis, P.G., Alexander, T.J., Krolikowski, W., Kivshar, Y.S.: Stable higher-charge discrete vortices in hexagonal optical lattices. Phys. Rev. A **79**, 025801 (2009)
30. Kartashov, Y.V., Ferrando, A., Egorov, A.A., Torner, L.: Soliton topology versus discrete symmetry in optical lattices. Phys. Rev. Lett. **95**, 123902 (2005)
31. Bekshaev, A.Y., Soskin, M.S., Vasnetsov, M.V.: Transformation of higher-order optical vortices upon focusing by an astigmatic lens. Opt. Commun. **241**, 237 (2004)
32. Terhalle, B., Richter, T., Law, K.J.H., Göries, D., Rose, P., Alexander, T.J., Kevrekidis, P.G., Desyatnikov, A.S., Krolikowski, W., Kaiser, F., Denz, C., Kivshar, Y.S.: Observation of double-charge discrete vortex solitons in hexagonal photonic lattices. Phys. Rev. A **79**, 043821 (2009)
33. Desyatnikov, A.S., Sagemerten, N., Fischer, R., Terhalle, B., Träger, D., Neshev, D.N., Dreischuh, A., Denz, C., Krolikowski, W., Kivshar, Y.S.: Two-dimensional self-trapped nonlinear photonic lattices. Opt. Express **14**, 2851 (2006)
34. Bezryadina, A., Neshev, D.N., Desyatnikov, A., Young, J., Chen, Z., Kivshar, Y.S.: Observation of topological transformations of optical vortices in two-dimensional photonic lattices. Opt. Express **14**, 8317 (2006)
35. Terhalle, B., Richter, T., Desyatnikov, A.S., Neshev, D.N., Krolikowski, W., Kaiser, F., Denz, C., Kivshar, Y.S.: Observation of multivortex solitons in photonic lattices. Phys. Rev. Lett. **101**, 013903 (2008)

Chapter 6
Summary and Outlook

Nonlinear photonic crystals offer great potential for controlling and manipulating the propagation of light and have consequently become an important research area of modern optics in recent years. In such structures, the periodicity leads to photonic band gaps which can be seen as optical analogs of the electronic band gaps known from solid state physics. Moreover, the nonlinear response of the material allows for dynamic tunability of structural parameters and thus the combination of both, nonlinearity and periodicity provides a unique opportunity to achieve an ultimate control over linear as well as nonlinear light propagation. As successful experiments in this context are strongly facilitated by reconfigurable refractive index structures of different symmetries which are easy to fabricate and possess strong nonlinearities, preferably at low laser powers, the optical induction in photorefractive media is used as a method to create the desired structures throughout this work. Due to their tunable and reconfigurable nature, these optically induced photonic lattices provide an ideal test bench for fundamental studies of wave propagation in nonlinear periodic systems.

The work presented in this thesis advances the knowledge of fundamental effects and experimental methods to control linear and nonlinear light propagation in such structures. Based on a detailed description of optically induced lattices in photorefractive media, it is revealed that, due to the photorefractive anisotropy, the induced refractive index change strongly depends on the spatial orientation of the lattice wave. As a key result, it is shown that, in the case of hexagonal lattices, the influence of anisotropy on the induced refractive index change can be compensated by stretching the lattice wave along its vertical direction. This way, a symmetric refractive index structure can be achieved and the experimental observation of fundamental phenomena associated with truly hexagonal structures becomes accessible in optically induced lattices.

Therefore, the stretched lattice is employed to investigate the fundamental phenomena of Rabi oscillations and interband transitions in hexagonal photonic structures. Based on an analytical description of resonant transitions at the high symmetry points of the Brillouin zone known as Landau-Zener-Majorana model,

B. Terhalle, *Controlling Light in Optically Induced Photonic Lattices*, Springer Theses, DOI: 10.1007/978-3-642-16647-1_6, © Springer-Verlag Berlin Heidelberg 2011

the first experimental observation of Bragg resonance induced Rabi oscillations as well as Landau-Zener tunneling in two-dimensional hexagonal structures is demonstrated. The experimental results are corroborated by numerical simulations using the anisotropic photorefractive model, revealing an excellent agreement between the theoretical predictions, numerical simulations and experiments.

Furthermore, the possibility of nonlinearity induced interband coupling in photonic lattices is studied experimentally. Besides the general interest in these fundamental effects, a strong dependence on the initial conditions for the excitation of Rabi oscillations and Landau-Zener tunneling suggests potential applications such as optical switching in Fourier space or controlled Bloch wave generation.

Subsequently, the possibilities to control beam propagation by nonlinear light localization are studied with emphasis on shaping the transverse energy flow in complex light fields by generating self-trapped phase singularites. While the photorefractive anisotropy generally prevents the existence of such vortex solitons in homogeneous media, a periodic refractive index modulation enables the generation of stable discrete vortex solitons in photorefractive materials. However, the stability requires a specific phase relation to balance the energy flow within the structure. In hexagonal lattices, the phase condition can be fulfilled by appropriate lattice stretching and thus the anisotropic manipulation of the lattice wave enables an experimental control over the stability of discrete vortex solitons.

Based on the satisfied phase condition in the stretched hexagonal lattice, it is demonstrated that ring-shaped discrete vortex solitons containing two identical phase singularities, i.e. having the topological charge 2, can be stable in the presence of a focusing nonlinearity while their single-charge counterparts are unstable. This is in strong contrast to homogeneous media in which a higher topological charge usually leads to stronger instabilities. In addition, it is shown that the stability properties are reversed in the case of a defocusing nonlinearity with the single-charge vortex appearing to be stable and the double-charge vortex exhibiting a weak instability.

Finally, the investigations are extended to more complex probe beam fields containing six phase singularities, leading to the observation of stable multivortex solitons in optically induced stretched hexagonal lattices. This is a particularly striking result as it provides the first evidence for the existence of stable multi-vortex clusters in the presence of attractive nonlinear interaction in any field of physics.

In connection with the work presented in this thesis, several directions for future studies can be imagined. Since all the presented results have been achieved in two-dimensional structures of hexagonal symmetry, the next step is to explore the features of more sophisticated lattice geometries for manipulating and controlling light propagation in photonic structures. Keeping the advantages in handling and flexibility provided by optically induced lattices, this implies the consideration of more complex nondiffracting waves and, as demonstrated in this thesis, a careful examination of the influence of anisotropy on the induced refractive index structures. As an example, one may consider the experimental realization of localized structures in kagome lattices which have recently been predicted theoretically [1]. The investigation of light propagation in complex lattice geometries may also establish additional links to the field of Bose-Einstein condensates in optical lattices where sophisticated trapping geometries have already been realized experimentally [2, 3].

Similarly, the implementation of three-dimensional periodic structures provides an exciting research direction with advanced features for linear as well as non-linear light propagation. While the optical induction of three-dimensional photonic lattices in photorefractive media has recently been demonstrated [4, 5], the corresponding propagation effects are yet to be explored.

Finally, the application of the presented concepts to smaller structures and subsequently the integration in photonic devices for future all-optical information processing is certainly another interesting direction to follow.

References

1. Law, K.J.H., Saxena, A., Kevrekidis, P.G., Bishop, A.R., et al.: Localized structures in kagome lattices. Phys. Rev. Lett. **96**, 180406 (2006)
2. Burkov, A.A., Demler, E.: Vortex-peierls states in optical lattices. Phs. Rev. Lett. **96**, 180406 (2006)
3. Santos, L., Baranov, M.A., Cirac, J.I., Everts, H.-U., Fehrmann, H., Lewenstein, M.: Atomic quantum gases in kagome lattices. Phys. Rev. Lett. **93**, 030601 (2004)
4. Xavier, J., Rose, P., Terhalle, B., Denz, C.: Three-dimensional optically induced quasicrystallographic three-dimensional complex nonlinear photonic lattice structures. Opt. Lett. **34**, 2625 (2009)
5. Xavier, J., Rose, M.B.P., Joseph, J., Denz, C.: Reconfigurable optically induced quasicrystallographic three-dimensional complex nonlinear photonic lattice structures. Adv. Mater. **22**, 356 (2010)

Chapter 7
Appendices

7.1 Numerical Methods

7.1.1 Solving the Potential Equation

In Sect. 2.2.4, we have derived the following equation for the electrostatic potential of the induced space charge field [cf. (2.37)]:

$$\nabla^2 \phi + \nabla \ln (1 + I) \nabla \phi = |E_0| \frac{\partial \ln (1 + I)}{\partial x}. \tag{7.1}$$

Numerical solutions to this equation can be found by an iterative procedure which has been suggested in [1]. The basic idea is to treat (7.1) as an inhomogeneous Poisson equation in which the mixed term on the left hand side is assumed to be a constant addition to the source term and is recalculated in each iteration step. Therefore, in order to stress the similarities to a Poisson equation, (7.1) is often rewritten as

$$\Delta \phi = U_0 - \nabla \ln (1 + I) \nabla \phi, \qquad U_0 = |E_0| \frac{\partial \ln (1 + I)}{\partial x}. \tag{7.2}$$

In the first step, the mixed term is set to zero such that the algorithm starts with

$$\Delta \phi_0 = U_0. \tag{7.3}$$

Transforming this equation into Fourier space yields

$$\mathcal{F}[\phi_0] = -\frac{\mathcal{F}[U_0]}{k_x^2 + k_y^2} \tag{7.4}$$

where k_x and k_y are the spatial frequencies and $\mathcal{F}$ denotes the Fourier transform. From this expression, ϕ_0 is obtained by an inverse Fourier transform and subsequently used in the iteration

B. Terhalle, *Controlling Light in Optically Induced Photonic Lattices*, Springer Theses, DOI: 10.1007/978-3-642-16647-1_7, © Springer-Verlag Berlin Heidelberg 2011

$$\Delta\phi_{i+1} = U_0 - \nabla \ln (1 + I)\nabla\phi_i. \tag{7.5}$$

An error estimate for the numerical calculation can be defined by the relative variation of the potential per iteration step

$$\Delta\xi_i = \frac{\sum_{k,l} |\phi_i(x_k, y_l) - \phi_{i-1}(x_k, y_l)|}{\sum_{k,l} |\phi_i(x_k, y_l)|}, \tag{7.6}$$

and the iteration is carried out until $\Delta\xi_i$ becomes smaller than a predefined lower bound. Throughout this work, we use $\Delta\xi_i < 10^{-5}$ as a stop criterion.

7.1.2 The Beam Propagation Method

In general, the propagation equation (3.1) as a nonlinear partial differential equation does not allow for analytical solutions and has to be solved numerically. Among various approaches, the split-step beam propagation method has been shown to be rather fast and reliable [2] and is therefore briefly reviewed in the following.

In this approach, an approximate solution to (3.1) is obtained by assuming that diffraction and nonlinear effects can be treated independently for a small propagation distance h. Under this assumption, the propagation from z to $z + h$ is calculated in two alternating steps in which either diffraction effects are considered and nonlinearity is neglected or vice versa (hence the notation as split-step method). To understand the basic procedure, it is convenient to rewrite (3.1) formally as

$$\frac{\partial A}{\partial z} = (\mathbf{D} + \mathbf{N})A \tag{7.7}$$

with two operators $\mathbf{D} = i\nabla_\perp^2/2$ and $\mathbf{N} = -i\Gamma E_{sc}/2$ accounting for diffraction and nonlinearity, respectively. A formal solution to this equation for the propagation from z to $z + h$ is given by

$$A(x, y, z + h) = e^{[h(\mathbf{D}+\mathbf{N})]}A(x, y, z). \tag{7.8}$$

Based on this expression, the numerical model is derived by applying the well-known Baker–Hausdorff formula [3]

$$e^{h\mathbf{D}} e^{h\mathbf{N}} = e^{h(\mathbf{D}+\mathbf{N})+\frac{1}{2}h^2[\mathbf{D},\mathbf{N}]+\cdots} \tag{7.9}$$

where $[\mathbf{D}, \mathbf{N}] = \mathbf{D}\mathbf{N} - \mathbf{N}\mathbf{D}$. Due to the noncommuting nature of the operators $\mathbf{D}$ and $\mathbf{N}$, a calculation of diffraction and nonlinear effects in two steps, i.e. replacing $e^{h(\mathbf{D}+\mathbf{N})}$ by $e^{h\mathbf{D}} e^{h\mathbf{N}}$ then results in a local accuracy of $\mathcal{O}(h^2)$ which turns out to be sufficient in most cases of interest. If necessary, the accuracy can also be improved by including higher order terms [2].

The diffraction is obtained by using a pseudospectral method including a Fourier transform of the envelope at the propagation distance z to facilitate the computation of the differential operator by $\nabla_\perp^2 \to -(k_x^2 + k_y^2)$. An inverse Fourier transform then gives the diffracted field envelope at the propagation distance $z + h$. The corresponding nonlinear correction is subsequently applied in real space such that the computation of the propagation can be summarized as

$$A(x,y,z+h) = e^{-i\Gamma E_{sc}h/2}\mathcal{F}^{-1}\left[e^{-i(k_x^2+k_y^2)h/2}\mathcal{F}[A(x,y,z)]\right]. \tag{7.10}$$

In our simulations, the Fourier transform is calculated using the freely available FFTW implementation of the discrete Fast Fourier Transform (FFT) [4]. Therefore, it should be noted that the discrete Fourier transform implies periodic boundary conditions which can lead to numerical instabilities since energy reaching one edge of the numerical grid automatically reenters from the other side. Although these effects can usually be neglected as long as the beam stays localized, it may cause problems if the propagation of larger beams or the interaction between several beams is considered. Apart from the very inefficient way of simply enlarging the grid, these problems can be circumvented by introducing absorbing boundary conditions. A suitable modification of the beam propagation method for this case can be found in [5].

7.1.3 Finding Solitary Solutions

The standard iterative procedure for the numerical calculation of solitons in nonlinear media was first suggested by Petviashvili in 1976 [6] and applied to homogeneous photorefractive media in 1998 [7].

As already discussed in Chap. 3, solitons in optically induced photonic lattices are found as solutions of (3.2) using the ansatz

$$A(r) = \psi(r_\perp) \cdot e^{i\beta z}. \tag{7.11}$$

Adopting the original algorithm for this case, the iteration scheme is therefore given by

$$\mathcal{F}[\psi_{n+1}] = |M|^{-3/2}\frac{\mathcal{F}\left[-\Gamma E_{sc}\left(|A_{latt}|^2 + |\psi_n|^2\right)\psi_n\right]}{2\beta + k_\perp^2} \tag{7.12}$$

with

$$M = \frac{\int d\mathbf{k}_\perp \mathcal{F}\left[-\Gamma E_{sc}\left(|A_{latt}|^2 + |\psi_n|^2\right)\psi_n\right](\mathcal{F}[\psi_n])^*}{\int d\mathbf{k}_\perp (2\beta + k_\perp^2)|\mathcal{F}[\psi_n]|^2}. \tag{7.13}$$

It has been shown that the factor $|M|^{-3/2}$ in (7.12) is necessary to ensure convergence of the iteration. If $\psi(\boldsymbol{r}_\perp)$ is solitary, M equals 1 and (7.12) becomes the Fourier transformed version of (3.2).

As before, the relative differences between two subsequent iterations can be used as a stop criterion:

$$\Delta\xi_n = \frac{\int d\boldsymbol{r}_\perp |\psi_n(\boldsymbol{r}_\perp) - \psi_{n-1}(\boldsymbol{r}_\perp)|^2}{\int d\boldsymbol{r}_\perp |\psi_n(\boldsymbol{r}_\perp)|^2}. \tag{7.14}$$

It is important to note that the above procedure is only applicable for solitons in the semi-infinite gap. In fact, this restriction even remains after further generalization of the original method [8]. A way to overcome these limitations is given by replacing the Fourier transform by a more appropriate Bloch transform. For a description of the corresponding algorithm, we refer to [5].

7.2 Characterization of the Spatial Light Modulator

In our experiments, we use the PLUTO-VIS phase only spatial light modulator from HOLOEYE Photonics AG. It is based on a reflective LCOS microdisplay with HDTV resolution (1920×1080 px) and can be addressed by an HDTV capable graphics card at a frame rate of 60 Hz. The display area is 18 mm $\times$ 10 mm and has a pixel size of $8 \times 8\,\mu$m. The effective phase shift $\Delta\phi$ results from a pulse-width modulation and can be achieved with two firmwares differing in the length of the puls sequences. The first firmware has a longer sequence which allows for addressing 1215 phase values. However, this sequence can only be sent twice per frame such that the phase varies at a frequency of 2×60 Hz $= 120$ Hz. The second firmware provides a shorter sequence which can be sent five times per frame leading to a frequency of 5×60 Hz $= 300$ Hz for the phase variation. However, this firmware enables only 192 different phase values. In both cases, the performance of the modulator may be optimized by a configuration file in which each gray value G can be assigned a particular phase value P.

To determine the effective phase shift $\Delta\phi$ experimentally, we use the setup shown in Fig. 7.1. The laser beam is expanded and passes through an aperture

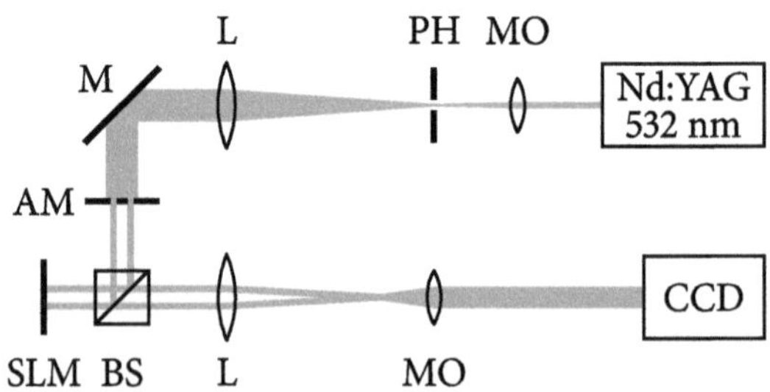

Fig. 7.1 Experimental setup for the characterization of the phase modulator. *AM* amplitude mask; *BS* beam splitter; *CCD* camera; *L* lens; *M* mirror; *MO* microscope objective; *PH* pinhole; *SLM* spatial light modulator

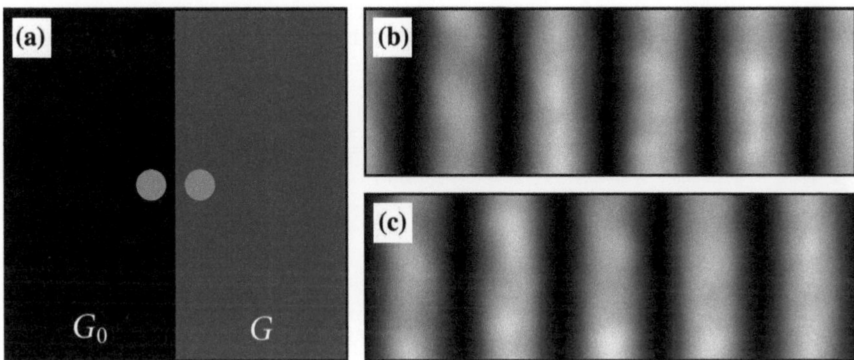

Fig. 7.2 Characterization of the phase modulator. **a** Schematic illustration of the illuminated area. One half is kept at a fixed gray value G_0 while the gray value G of the other half is varied; **b**, **c** typical interference patterns for a relative phase difference of zero (*top*) and π (*bottom*) between the two beams

plate in order to generate two coherent beams which are subsequently sent onto the modulator. The two beams are reflected from the modulator and interfere in the focal plane of a lens. The resulting interference pattern is finally imaged onto a CCD camera using a microscope objective.

As shown in Fig. 7.2a, the beams illuminate different halves of the display area where one half is kept at a fixed gray value G_0 while the gray value of the other half is varied. This variation leads to a phase shift of the corresponding beam which results in a spatial shift of the interference pattern [Fig. 7.2b and c].

Analyzing the relative spatial shift of the interference patterns for different gray values therefore enables the determination of the phase characteristics. The actual measurement is performed by taking several images and determining the average position of the maxima and minima in the interference pattern for each phase value. The effective phase shift $\Delta\phi$ is obtained by comparing the positions of the

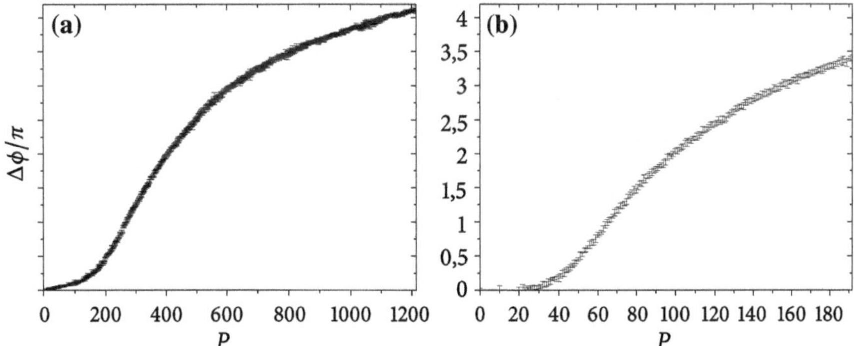

Fig. 7.3 Experimentally obtained phase shift $\Delta\phi$ as a function of the available phase values. **a** Firmware 1; **b** firmware 2

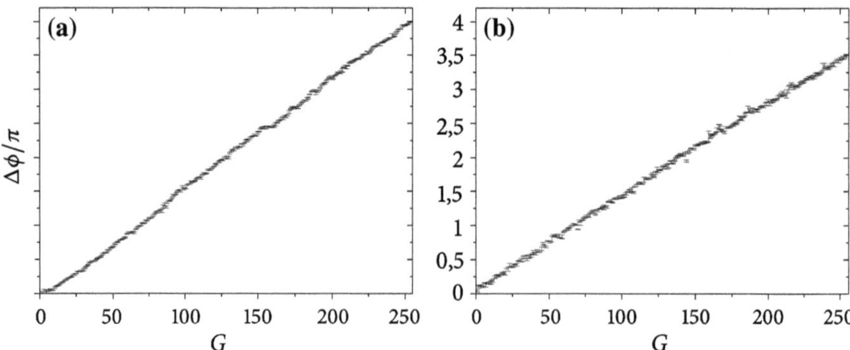

Fig. 7.4 Phase characteristics with a configuration file optimized for a linear dependence of the phase shift $\Delta\phi$ on the gray value G. **a** Firmware 1; **b** firmware 2

extrema for each phase value P with the original positions for $P = 0$. The result is shown in Fig. 7.3 for both firmware versions. Note that, as a single measurement only allows for addressing 256 phase values, the calibration curve for firmware 1 (Fig. 7.3a) is the result of several subsequent measurements.

In most experiments, it is convenient to have a linear dependence of the phase shift $\Delta\phi$ on the gray value G. Therefore, a suitable configuration file has been generated for both firmware versions (Fig. 7.4). This way, a very good linear dependence could be achieved in both cases.

7.3 The Phase Imprinting Technique

In order to employ the spatial light modulator for the generation of complex probe beams consisting of several intensity spots with well defined phases, the phase imprinting technique has been used. This technique was first suggested in [9] for selective excitation of Bloch waves in optically induced photonic lattices and has been further optimized in the framework of this thesis. Although the used Holoeye PLUTO-VIS modulator is originally a phase only device, the phase imprinting technique allows for an almost independent experimental control of the probe beam intensity and phase structure.

The basic concept is illustrated in Fig. 7.5 for a simple Gaussian intensity distribution with a superimposed binary phase structure.

We use a periodic grating with a slanted cross-sectional profile, known as Blaze grating, which makes it possible to concentrate most of the energy in the first diffraction order. This Blaze grating is then multiplied with the aimed spatial intensity distribution (normalized to 1). Hence, only the regions of the modulator in which the intensity distribution is not equal to zero will be covered by the Blaze grating and consequently only light from these regions will be directed into the first diffraction order.

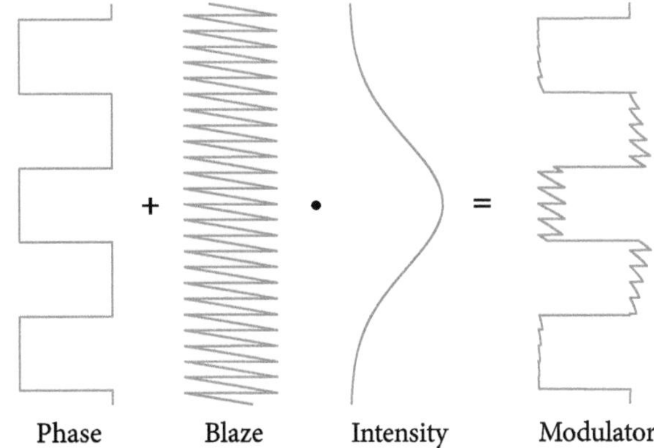

Fig. 7.5 Schematic illustration of the phase imprinting technique for a Gaussian intensity distribution with a binary phase structure

The desired phase distribution is simply added such that the diffracted light also carries the phase information. Proper Fourier filtering in the focal plane of the telescope following the modulator (cf. Fig. 5.5.) finally ensures that only light with the desired intensity and phase distribution is sent onto the crystal.

As an example, Fig. 7.6 shows the implementation of the phase imprinting technique for the generation of three-lobe clusters with unit topological charge as studied in Sect. 5.2. The blaze grating shown in Fig. 7.6a is multiplied with an amplitude mask which contains the three lobes of the final beam as circles [Fig. 7.6b]. Within these circles, the corresponding matrix is set to one while the region outside these circles is zero. In order to obtain the required vortex structure, the relative phase differences between the lobes are $\Phi_{ji} = 2\pi/3$ as demonstrated in the phase profile shown in Fig. 7.6c. The final matrix is shown in Fig. 7.6d. As described above, it is obtained by multiplying the Blaze grating with the amplitude mask and subsequently adding the phase information.

By giving this matrix onto the modulator and illuminating it with a broad plane wave, only light from the three lobes will be diffracted into the first order, while the black area simply reflects the incomming wave into the zeroth order. As the diffracted light also carries the phase information, filtering out the first order

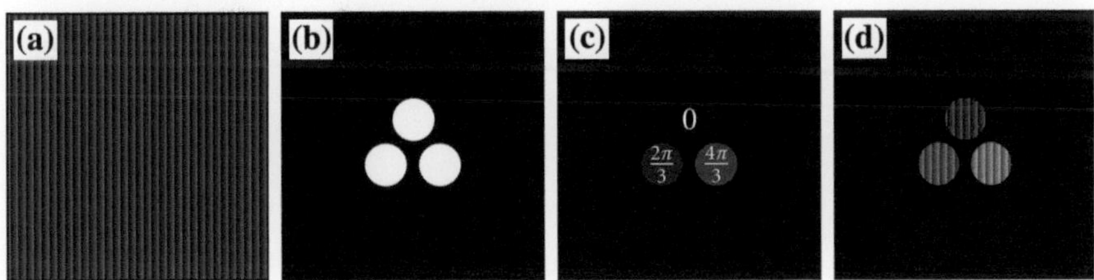

Fig. 7.6 Generation of a three-lobe input beam with topological charge $M = +1$ using the phaseimprinting technique. **a** Blaze grating; **b** amplitude mask; **c** phase structure; **d** resulting matrix given onto the modulator

results in the desired probe beam structure of three lobes with a phase profile of unit topological charge shown in Fig. 5.6a and d.

References

1. Stepken, A.: Optische räumliche Solitonen in photorefraktiven Kristallen, Dissertation (TU Darmstadt), Shaker Verlag, Aachen (2001)
2. Agrawal, G.P.: Nonlinear Fiber Optics (2nd edn.) Academic press, San Diego (1995)
3. Weiss, G.H., Maradudin, A.A.: The Baker-Hausdorff formula and a problem in crystal physics. J. Math. Phys. **3**, 771–777 (1962)
4. Frigo, M., Johnson, S.G.: The design and implementation of FFTW3. In: Proceedings of the IEEE 93 (2005)
5. Richter, T.: Stability of anisotropic gap solitons in photorefractive media. PhD thesis, TU Darmstadt (2008)
6. Petviashvili, V.I.: On the equation of a nonuniform soliton. Fiz. Plasmy **2**, 469–472 (1976)
7. Zozulya, A.A., Anderson, D.Z., Mamaev, A.V., Saffman, M.: Solitary attractors and low-order filamentation in anisotropic self-focusing media. Phys. Rev. A **57** (1998)
8. Musslimani, Z.H., Yang, J.: Self-trapping of light in a two-dimensional photonic lattice. J. Opt. Soc. Am. B **21**, 973–981 (2004)
9. Träger, D., Fischer, R., Neshev, D.N., Sukhorukov, A.A., Denz, C., Krolikowski, W., Kivshar, Y.S.: Nonlinear Bloch modes in two-dimensional photonic lattices. Opt. Express **14**, 1913–1923 (2006)

Curriculum Vitae

Personal Data

Name: Bernd Terhalle

Date of Birth: May 6, 1981

Place of Birth: Haselünne

Citizenship: German

Marital Status: Unmarried

School Education

1987–1991   Grundschule Börger, Germany

1991–1993   Orientierungsstufe Sögel, Germany

1993–2000   Hümmling Gymnasium Sögel, Germany

July 2000   Final secondary school examinations

Civilian Service

2000–2001   Malteser Hilfsdienst Esterwegen, Germany

Higher Education

2001–2003	Basic study period in physics, Westfälische Wilhelms-Universität Münster, Germany
Sept. 2003	Intermediate diploma
2003–2004	Advanced study period in physics, Westfälische Wilhelms-Universität Münster, Germany
2004–2005	Study abroad, James Cook University Townsville, Australia
2005–2006	Diploma thesis: *"Anisotropic photonic lattices and discrete solitons in photorefractive media"*, Institut für Angewandte Physik, Westfälische Wilhelms-Universität Münster, Germany
Jan 2007	Diploma
Since Feb. 2007	PhD studies, Institut für Angewandte Physik, Westfälische Wilhelms-Universität Münster, Germany

Honors and Awards

July 2007	Infineon Master Award for the best diploma thesis in physics, Westfälische Wilhelms-Universität Münster, Germany
2007–2010	DAAD graduate scholarship for a binational PhD project

International Experience

2004–2005	Study abroad, James Cook University Townsville, Australia
2007–2009	Annual research visits, Australian National University, Canberra, Australia

Work Experience

2003–2006   Tutoring student assistant, Mathematisches Institut, Westfälische Wilhelms-Universität Münster, Germany

Since 2007   Research assistant, Institut für Angewandte Physik, Westfälische Wilhelms-Universität Münster, Germany

Publications

Journal Articles

- **B. Terhalle**, D. Göries, T. Richter, P. Rose, A. S. Desyatnikov, F. Kaiser, and C. Denz,*Anisotropy-controlled topological stability of discrete vortex solitons in optically induced photonic lattices*, Opt. Lett. **35**, 604 (2010).

- C. Denz, **B. Terhalle**, P. Rose, J. Xavier, T. Richter, A. S. Desyatnikov, J. Imbrock, T. J. Alexander, J. Joseph, F. Kaiser, W. Krolikowski, and Y. S. Kivshar, *Nonlinear photonics in complex optically-induced photonic lattices*, Proc. SPIE **7354**, 735402 (2009).

- J. Xavier, P. Rose, **B. Terhalle**, J. Joseph, and C. Denz, *Three-dimensional optically induced reconfigurable photorefractive nonlinear photonic lattices*, Opt. Lett. **34**, 2625 (2009).

- **B. Terhalle**, T. Richter, K. J. Law, D. Göries, P. Rose, T. J. Alexander, P. Kevrekidis, A. S. Desyatnikov, W. Krolikowski, F. Kaiser, C. Denz, and Y. S. Kivshar, *Observation of double charge discrete vortex solitons in hexagonal photonic lattices*, Phys. Rev. A **79**, 043821 (2009).

- P. Rose, **B. Terhalle**, J. Imbrock, and C. Denz, *Optically-induced photonic superlattices by holographic multiplexing*, J. Phys. D: Appl. Phys. **41**, 224004 (2008).

- **B. Terhalle**, T. Richter, A. S. Desyatnikov, D. N. Neshev, W. Krolikowski, F. Kaiser, C. Denz, and Y. S. Kivshar, *Observation of multivortex solitons in photonic lattices*, Phys. Rev. Lett. **101**, 013903 (2008).

- **B. Terhalle**, N. Radwell, P. Rose, C. Denz, and T. Ackemann, *Control of broad-area vertical-cavity surface emitting laser emission by optically induced photonic crystals*, Appl. Phys. Lett. **93**, 151114 (2008).

- P. Rose, T. Richter, **B. Terhalle**, J. Imbrock, F. Kaiser, and C. Denz, *Discrete and dipole-mode gap solitons in higher-order nonlinear photonic lattices*, Appl. Phys. B **89**, 521 (2007).

- **B. Terhalle**, A. S. Desyatnikov, C. Bersch, D. Träger, L. Tang, J. Imbrock, Y. S. Kivshar, and C. Denz, *Anisotropic photonic lattices and discrete solitons in photorefractive media*, Appl. Phys. B **86**, 399 (2007).

- **B. Terhalle**, D. Träger, L. Tang, J. Imbrock, and C. Denz, *Structure analysis of two-dimensional nonlinear self-trapped photonic lattices in anisotropic photorefractive media*, Phys. Rev. E **74**, 057601 (2006).

- A. S. Desyatnikov, N. Sagemerten, R. Fischer, **B. Terhalle**, D. Träger, D. N. Neshev, A. Dreischuh, C. Denz, W. Krolikowski, and Y. S. Kivshar, *Two-dimensional self-trapped nonlinear photonic lattices*, Opt. Express **14**, 2851 (2006).

Book Chapters

- J. Imbrock, **B. Terhalle**, P. Rose, P. Jander, S. Koke, and C. Denz, *Complex nonlinear photonic lattices: from instabilities to control* in Nonlinearities in Periodic Structures and Metamaterials (Eds. C. Denz, S. Flach, and Y. S. Kivshar), Springer Verlag, Berlin (2010), ISBN 978-3-642-02065-0.

- **B. Terhalle**, P. Rose, D. Göries, J. Imbrock, and C. Denz, *Dynamics and nonlinear light propagation in complex photonic lattices* in Nonlinear Dynamics of Nanosystems (Eds. G. Radons, B. Rumpf, and H. G. Schuster), Wiley-VCH Verlag, Weinheim (2010), ISBN 978-3-527-40791-0.

Selected Conference Contributions

- P. Rose, M. Boguslawski, **B. Terhalle**, J. Imbrock, and C. Denz, *Complex photonic superlattices optically induced in nonlinear media*, ICO Topical Meeting on Emerging trends and Novel Materials in Photonics, Delphi (2009).

- **B. Terhalle**, A. S. Desyatnikov, D. N. Neshev, W. Krolikowski, C. Denz, and Y. S. Kivshar, *Landau-Zener tunneling dynamics in hexagonal photonic lattices*, CLEO/Europe-EQEC, Munich (2009).

- D. Göries, **B. Terhalle**, P. Rose, T. Richter, T. J. Alexander, A. S. Desyatnikov, W. Krolikowski, F. Kaiser, Y. S. Kivshar, and C. Denz, *Observation of double charge discrete vortex solitons in hexagonal photonic lattices*, Photorefractive Materials, Effects and Devices, Bad Honnef (2009).

- **B. Terhalle**, A. S. Desyatnikov, D. N. Neshev, W. Krolikowski, C. Denz, and Y. S. Kivshar, *Observation of Landau-Zener tunneling in hexagonal photonic lattices*, Photorefractive Materials, Effects and Devices, Bad Honnef (2009).

- **B. Terhalle**, C. Denz, A. S. Desyatnikov, T. Richter, D. Göries, P. Rose, T. J. Alexander, D. N. Neshev, F. Kaiser, W. Krolikowski, and Y. S. Kivshar, *Vortex clusters in photonic lattices* (invited talk), IEEE LEOS-Winter Topicals, Innsbruck (2009).

- **B. Terhalle**, A. S. Desyatnikov, T. Richter, T. J. Alexander, D. N. Neshev, F. Kaiser, C. Denz, W. Krolikowski, and Y. S. Kivshar, *Multivortex solitons in photonic lattices*, Australian Institute of Physics Congress, Adelaide (2008).

- P. Rose, **B. Terhalle**, T. Richter, J. Imbrock, F. Kaiser, and C. Denz, *Light propagation in complex two-dimensional photonic lattices*, Discrete Optics and Beyond Workshop, Bad Honnef (2008).

- **B. Terhalle**, T. Richter, A. S. Desyatnikov, D. N. Neshev, W. Krolikowski, F. Kaiser, C. Denz, and Y. S. Kivshar, *Observation of multivortex solitons in photonic lattices*, Discrete Optics and Beyond Workshop, Bad Honnef (2008).

- N. Radwell, **B. Terhalle**, P. Rose, C. Denz, and T. Ackemann, *Control of broad-area vertical-cavity surface emitting laser emission by optically induced photonic crystals*, Annual meeting of the European Optical Society, Paris (2008).

- P. Rose, **B. Terhalle**, J. Imbrock, and C. Denz, *Multiplexing photonic superlattices in photorefractive media*, Annual meeting of the European Optical Society, Paris (2008).

- **B. Terhalle**, P. Rose, T. Richter, A. S. Desyatnikov, C. Bersch, F. Kaiser, Y. S. Kivshar, and C. Denz, *Aspects of anisotropy in nonlinear photonic lattices*, Nonlinear Optics Applications Workshop, Swinouscie (2007).

- P. Rose, **B. Terhalle**, T. Richter, A. S. Desyatnikov, C. Bersch, J. Imbrock, F. Kaiser, Y. S. Kivshar and C. Denz, *Anisotropic self-focusing in two-dimensional photonic lattices*, Spring meeting of the German Physical Society, Düsseldorf (2007).

- **B. Terhalle**, D. Träger, A. S. Desyatnikov, C. Bersch, J. Imbrock, R. Fischer, D. N.Neshev, A. A. Sukhorukov, W. Krolikowski, Y. S. Kivshar, and C. Denz, *Nonlinear wave propagation and photonic lattices in anisotropic photorefractive media*, Nonlinear Dynamics of Nanosystems Symposium, Chemnitz (2006).

- **B. Terhalle**, A. S. Desyatnikov, C. Bersch, D. Träger, L. Tang, J. Imbrock, Y. S. Kivshar, and C. Denz, *Anisotropic photonic lattices and discrete solitons in photorefractive media*, Annual meeting of the European Optical Society, Paris (2006).

- **B. Terhalle**, A. S. Desyatnikov, D. Träger, L. Tang, D. N. Neshev, W. Krolikowski, Y. S. Kivshar, and C. Denz, *Two-dimensional self-trapped photonic lattices in anisotropic photorefractive media*, Spring meeting of the German Physical Society, Frankfurt (2006).

MIX
Papier aus verantwortungsvollen Quellen
Paper from responsible sources
FSC® C105338

www.fsc.org

If you have any concerns about our products,
you can contact us on
ProductSafety@springernature.com

In case Publisher is established outside the EU,
the EU authorized representative is:
Springer Nature Customer Service Center GmbH
Europaplatz 3, 69115 Heidelberg, Germany

Printed by Libri Plureos GmbH
in Hamburg, Germany